Before Patty
Volume 1

Also by the Author:

Iskut Ridge (fiction)

The Monk Who Howled Like a Wolf: The Mystic's Path of Kriya Yoga (nonfiction)

A statistical analysis of 130 trailside twig breaks in a New England woodland identifies 16.2% unlikely to have been created by H. sapiens or local fauna [pre-print paper]

Before Patty
Volume 1

Patrick—the Sasquatch/Human Hybrid & Our Genetic Inheritance

Norman Anton Sollie

Iola Publishing
Healy, Alaska

Iola Publishing
P.O. Box 560
Healy, Alaska USA

Cover design by the author.

Printed by KDP

ISBN: 978-1-948189-10-1 (hardcover edition)

DEDICATION

To Mary Dawn Wilson, whose July 12, 2022 distressing disappearance on the Stampede Road in Interior Alaska remains a mystery.

ACKNOWLEDGEMENTS

My first acknowledgment goes to the many scientists, researchers, and authors contributing to this work as primary sources, noted and credited in text, footnotes, and bibliography.

I would also like to thank Larry "Beans" Baxter (author and producer of the Alasquatch podcast); Jessie Desmond (anthropological research), Boreal Bigfoot Research Group (BBRG); Mia Feroleto; Monty Freeman; Wes Guermer and *Sasquatch Chronicles*; Heather Moser (genealogy research), author and film producer; Taylor Lanahan; Cass Ray; Shellie Rosan; Douglas Shepherd, wildlife biologist and BBRG; Maria Shepherd, wildlife biologist; Michael Thompson, SasquatchTracker.com and the BBRG; and Heidi Worley, BBRG.

CONTENTS

ONE

The Bride Purchase

"These Indian devils would sometimes watch the camps of the Indians very closely and follow them about as they moved from place to place, watching for an opportunity to seize one of the young women and carry her off to make her his wife…. The women of the tribe had great fear of them…. Sometimes the women would be captured by the Indian devils and be gone away from their tribe for years, when they would return and tell of their wild life and experiences. They would become the mother of children, and the children would inherit the wild habits of their father…"

---Lucy Thompson 1991
(original ed. 1916) pages 129-130

In 1891—seventy-five years before the October 1967 Patterson-Gimlin film of a female Sasquatch broke upon the world—a petite sixteen-year-old Native American woman stopped her walk through the rust and green ponderosa pines and set down a large, battered metal pot on the soft dry ground. She wrapped her canvas jacket more tightly around thin shoulders. The sudden chill that came over her wasn't from the incessant wind in this broad, but warm valley. The feeling of being watched had followed her for days at this fish camp on the Sanpoil River.

Hahissiat Kolockun,[1] was born of one of the twelve Confederated tribes. In 1872 these tribes were confined onto this still teenage reservation. Hahissiat was sixteen years old now—a new "bride-purchase." She had been sold out of her own tribe for smoked salmon, some horses, and trade goods to a man just a handful of years older than herself. But her arranged marriage luck was good—Pierre Paul had treated her well in the couple of weeks they had been together. And now this bustling salmon camp, with its groups of women gossiping over the smudge fires at the drying racks and the clusters of excited gamblers at the Stick games, promised abundance and a happy winter for the many families gathered here.

Described as "Colville woman, ca 1900-1910" by Latham Edward H." [**Author's note**: Dr E. H. Latham was the Colville Agency physician through about 1892 so this picture of a woman in traditional dress probably actually dates from the late nineteenth century.]

Although the numbers of salmon, eel, whitefish, suckers, and trout had been dropping in recent years—blamed on the hungry whites downriver—it was only mid-summer now and the catch had been good so far, with the late summer peak still ahead. Though the camp was small, it was filled with members of several of

[1] Patrick's mother's maiden name from the 10/30/1912, Patrick [redacted] Certificate of Birth; Delayed Birth Registration; Division of Vital Statistics; Washington State Department of Health

the twelve tribes and even had a couple tipis pitched by San Poil tribe traditionalists.

Smaller fish were caught with traps and long-handled dip nets; the larger salmon were speared. Cured fish would be traded all the way from the white invader cities along the coast to the growing settlement of Spokane to the east, and to mining and logging camps to the north. In return, flour and tools, rifles and cartridges, canned goods and building supplies would flow back to the reservation.

As long as the Columbia and its tributaries flowed free, the Twelve Tribes would prosper with the fish they caught and the grain crops they were starting to grow for their own horses and for the ever hungry white settlements.

Hahissiat stood and lifted one thigh to push the large pot more firmly up into her two-handed grip. Her mother-in-law, *Nanuquiya*[2] had sent Hahissiat for water to begin the evening meal. As she had been instructed, Hahissiat was climbing the short trail past the last campsite and the latrine area to draw water from the clean feeder creek dropping crystal clear across the rocks and then through the fish camp into the gentle green Sanpoil below.

She was almost to the water spot now, the path threading into shade among the increasingly thick trees and willow brush along the cool watercourse. She would be glad when she was back in the camp below and could shake this crawling feeling. Her new people were kind. Every Indian in the largely Moses-Columbia band united in their shared frustration with the increasing numbers of ever-hungry white settlers trespassing and squatting on their land.

Hahissiat was at the spot now—a rocky opening in the creek-side brush where the late afternoon sun was able to reach the soft ground. Her uneasiness faded as she set the pot down on her right next to a clear pool in the rocks. With her left hand, she gathered her long skirt to her side so that she could kneel at the water's edge, the basalt rock hard beneath her steadying right arm.

She hesitated for a moment before turning to reach for the

---

[2] Patrick's mother-in-law's name from the 11/23/1933 Pierre Paul [redacted] Certificate of Death; Bureau of Vital Statistics, Washington State Board of Health

container. Had she heard something from the brush? For a moment the sunlight seemed to grow dimmer somehow and she noticed that it was very quiet. Even the gurgling brook seemed to pause and hold its breath. One of the sacred dark-blue camp birds in the willows and cottonwoods across the creek broke the silence with a raucous screeching.

This sense of threat must just be her imagination. Life was not easy for Indians, particularly small as she was, but even with alcohol in the camp, rape was very unlikely for a married woman. Such a man would be driven away, disgraced by all, if he was able to escape Pierre Paul's anger.

The sooner she got out of this evil spot the better. She twisted and reached with both hands for the huge cooking pot and was lifting it up when she had a split second awareness of a *real* shadow engulfing her an instant before she was grabbed from behind. The pot flew through the air and landed with a great clattering as it bounced and tumbled down the rocks below her.

Hahissiat was lifted into the air by one massive arm around her chest. She began screaming as loud as she could with the breath rapidly being squeezed out of her. She kicked furiously at the solid body behind until she realized with a shock that what she thought was a torso was actually a muscular leg bigger than she imagined possible on any man.

She heard shouts in the camp below her and noticed that she was still screaming. What kind of Indian was this that carried her in an iron grip with just one arm, now pushing through the whipping brush and turning uphill?

He was enormous!

Hahissiat twisted downhill to look around the body of her abductor, hoping for help. Her view was mostly blocked by his massive side. It registered that this strange man must be dressed in hides, like the Indians before the time of the Whites, because she could see thick dark-brown hair. Leaning further back she saw two or three men below sprinting uphill through the trees. One man carried a rifle. Hahissiat began beating at the arm across her chest. The limb was like rock with muscle, as big as the biggest leg. She grabbed onto brush, trying to slow the rapid climb up the hill, limbs cutting her hands as the branches were ripped away from her.

She caught a glimpse of the man with the rifle lowering the barrel, his face full of horror.

"No, no, no!" she cried. "Don't let him take me!"

As her abductor climbed higher, covering ground with steps longer than she was tall, and the men below dropped away, she looked at the arm across her chest. Though human-like in form, the arm was not the arm of any Indian she had ever seen. The arm was *not* covered in animal skins. The arm was bare of sleeves, but covered in the thick, long hair that grew from his ash-gray skin.

This was no Indian man. This was one of the feared Forest People. Hahissiat was being carried off by a Stick Indian: *Skanicum*. She went limp as horror suddenly drained her strength.

Before Patty

CHAPTER TWO

The Return of Patrick

In 2019, Dr. Igor Burtsev leaned in close to me and whispered, "For *you*… two hoondred dollah."

Minutes before, I had listened intently to this seventy-something Russian hominid researcher speak in broken English of the long history of hairy biped sightings and encounters in Russia. Russel Accord, the 2019 International Bigfoot Conference organizer, had stood nearby to offer English words and quick repeats of what the often difficult-to-understand Dr. Burtsev was saying. Burtsev described the depth of encounters in Russia with beings seemingly identical to "our" Sasquatch and that there had been serious study of the phenomena there for decades. Dr. Burtsev had even done early size and motion work on the 1967 Patterson-Gimlin film.

When he concluded his presentation with an invitation to the audience to buy his book, I hurried to the vendor area, as a break began, and found a conference employee at Burtsev's table—I assumed he was a loaner helper from the soft-hearted Accord. There were a couple stacks of his book on the table edge and I was able to slip to the front of the milling scattering of attendees and carefully picked up one of the glossy softcover paperbacks. *Catching up with BIGFOOT: Kuzbass (Siberia) and Beyond…* The book was large, about 8 x 11-1/2" (20 x 29 cm) and I imagined him dragging the heavy volumes all the way from overseas in his luggage. Perhaps he was relying on book sales to defray the costs of his trip to North America. As a casual author myself, I was immediately sympathetic to his cause and also intrigued by the book.

The unique format of the book immediately caught my eye as I gently leafed through the precious volume. Each large page was divided into two columns. The first column was filled with text in the Cyrillic alphabet—presumably in Russian. The second column was in an English of uneven translation expertise. The entire book is filled with black and white and color photographs. It was obviously a very expensive book to print. I braced myself for the cost and mentally estimated that it would sell for at least $50.00, but there were no prices posted at the table or on the book that I held.

I asked the young man, "How much for the book?" I could feel the attention of the people around me—they wanted to know too.

"Oh!… I don't know. I was just supposed to keep an eye on the table while he was on stage."

Before I could react to this strange impasse, Dr. Burtsev himself walked up. The table sitter told the good doctor that I was asking about the price. The Russian doctor immediately responded loudly for all to hear in his thick accent, "Three hoondred dollah."

As I involuntarily gave a slight start, a woman behind me spit out, "Oh, I can't pay *that much* for a *book*!" as she turned and hustled away.

Dr. Burtsev took in the other dismayed reactions of the potential customers and stepped closer to me, leaning in.

"Too much? For *you*… two hoondred dollah."

What was I to do? This was Dr. Igor Burtsev, Director of the International Center of Hominology who had schlepped this one-of-a-kind book all the way from Moscow.

I pulled out my wallet and gave him cash.

*

Amused as I am by my Igor anecdote, I included it here because the book, *Catching Up with BIGFOOT*, plays a huge part in this story of Patrick.

After carefully transporting the book with me back to Alaska (returning the book three quarters of the way to the Siberia of its title), I skimmed through the tome a couple of times and looked at a lot of the fascinating pictures before setting the over 400 pages aside for when I had time for a careful read.

It wasn't until 2022 that I began working my way through this fascinating book. Inside I found an article from 2005 by Dmitri Bayanov, also of Moscow, entitled, "Is Manimal More Man than Animal?"[3] Within this article, Bayanov mentioned a 1992 paper by anthropologist, Dr. Ed Fusch: *S'cwene'yti and the Stick Indians of the Colvilles: The Interaction of Large Bipedal Hominids with American Indians as reported to Dr. Ed Fusch, Anthropologist, 1992.*[4]

Ironically, I had to haggle with a Russian researcher in Washington state to later read in Alaska about a foundational account of the Sasquatch world. Bayanov concluded the story in his article with, "…I wish our North American colleagues would devote as much time and effort to verifying the Patrick legend as Igor Bourtsev [sic] has devoted to exploring the Zana legend."

Zana was reported to be a wild woman, captured in Russia sometime in the latter decades of the 1800s and thought to have born children by more than one man in the village of her captivity.[5] Dr. Burtsev himself has gone to incredible lengths to attempt to establish the facts of Zana's story.

On November 19, 2022 I sent an email out to my Boreal Bigfoot Research Group (the BBRG—at that time consisting of Larry "Beans" Baxter, Jessie Desmond, Michael Thompson, and Heidi Worley) mentioning the claimed existence of a Sasquatch/Human hybrid named Patrick and linking the text of the original paper. I also copied Heather Moser of the

---

[3] Burtsev 2015 pages 305-307

[4] Dr. Jeff Meldrum, Bobbie Short, and David Paulides have mentioned this article/paper, along with several other researchers. I have also run into references to a videotaped interview with Dr. Fusch and Rhettman A. Mullis Jr., but have been unable to track the footage down. You may find the original paper without illustrations reproduced here:
http://www.bigfootencounters.com/biology/fusch.htm

[5] Debenat 2009 pages 203-205

documentary company Small Town Monsters. I had recently met Moser when she was part of a film crew getting documentary footage for productions on the Alaskan Sasquatch.

Moser needed no prodding—she is a talented and curious researcher. She immediately began investigating the genealogical hints dropped in the Fusch paper. On November 24th, Moser texted that she thought she had tracked down Patrick in census records, but that there were inconsistencies between the 1992 paper account and the public records she could access. We arranged for her to give a report to the BBRG, once she had a few more days to research.

In the meantime, I was looking in to Dr. Ed Fusch. On the 25th, I sent Moser my notes on Fusch. The next day Moser reported back that the last known residence that she could find for Fusch was in Havasu, Arizona. She had also tracked down a phone number for Ed. She tried the number several times, but no one ever picked-up and there was no voice mail.

On the 29th, Moser gave her presentation to the BBRG on what she was able to find out regarding Patrick's lineage and Dr. Fusch's whereabouts. Out of concern for living relatives, Moser did not reveal last names and specific dates to me or to the rest of the BBRG.

Based on what Moser and I had found so far, I believed that it was justified to put more research time into the possibility that an abduction in the late 1800s had led to a Sasquatch/Human hybrid child named Patrick. I began to research hybrids and the Colville Reservation. Moser and I, and eventually Desmond and Worley as well, talked about producing a book on the subject. Unfortunately, the others became far too busy with other projects, while my life had become simpler. Eventually, I decided to dig into public records and newspaper accounts myself to see if the evidence warranted our suspicion that there was something solid to this story. If so, I felt that Patrick's paradigm-shaking existence should be understood and shared.

Consider for a moment the curious serendipity of this child's name: *Patrick.* The name is not some cutesy nod to Patty, the name popularly bestowed on the female Sasquatch subject of the 1967 Patterson-Gimlin film. Patrick *was his name* and it appears in census records, marriage licenses, in allotment paperwork, newspaper articles, death certificates, obituaries, draft registration cards, and police records. What name could be more perfect for this genetic marvel—a hybrid between *H. sapiens* and Skanicum?

Skanicum and *S'cwene'yti* are both Native American names used for Sasquatch on the Colville Reservation.

Dr. Fusch's *S'cwene'yti and the Stick Indians of the Colvilles* was the starting place, of course. Dr. Fusch's info on Patrick was gleaned entirely from informant interviews on the Colville reservation and is anecdotal and so extremely subject to all the vagaries of perception bias, memory, familial needs, preconceptions, and imagination in both the informants and in Dr. Fusch. Moser's groundbreaking work had begun the long and painstaking process of finding documented facts to match or disprove the anecdotes.

Specific points the paper[6] alleges that may be tested through research:

> Interviews with the Colville tribal members (both women of the Lake Band) were conducted in 1985.

> Informant #1 was the woman "Francis," age 60 at the time of the interview (born about 1925).

> Informant #2 was the woman "Laura," mother to Francis, residing in Nespelem on the reservation, age 85 at the time of the interview in 1985; dying in 1987 at about age 87 Laura said she was niece to Patrick.[7]

> The abduction was "around the turn of the century (1885-1900)" at a fishing camp near Keller, WA.[8]

> The abductee was, "with Skanicum all summer, or at least a couple of months."

[6] Fusch 1992a pages 19-23

[7] It is common among North American First Nations to call even unrelated people of the older generation uncle, aunt, grandfather, or grandmother. This stated relationship tie with Patrick was not necessarily literal.

[8] Keller, in the Sanpoil river valley, was not established until a few years after the abduction, in 1898. The town was later moved more than once upstream due to rising reservoir waters before settling in its current spot, about ten miles southeast of Nespelem, in 1940.

> "During her stay with Skanicum the woman became pregnant and bore a son named Patrick, who grew up on the reservation."
> "Patrick's body structure was very different from that of other Indians":
  - Arms reaching "about to his knees"
  - 5'-4" tall
  - Sloping forehead
  - Large lower jaw
  - Large and wide mouth
  - Ears elongated upwards ("peaked") and bent out at the top
  - Large hands and long fingers
  - Very ugly
  - Louie [see below] described Patrick as a "pinhead" with larger than normal ears and large hands
> Patrick's personal and behavioral characteristics:
  - Extremely intelligent
  - A gentle man—never beat or mistreated his wife
> Biographical information on Patrick:
  - Operated a ranch in the area
  - Patrick married easily (to Laura's cousin) as he was considered "affluent"
  - Patrick had three daughters and two sons
  - Died at about age 30
  - Louie [see below] said that Patrick was a very good card player that seemed to always know what everyone else was holding
> Patrick's children:
  - Both unnamed sons died at an early age
  - Mary Louise (oldest birthdate daughter), about 65 years old in 1985 (so born about 1920) and living near the reservation. "Mary had heard several times over the years that her paternal grandfather was a Skanicum and sought verification from Laura [Laura about 20 years older than Mary Louise]. Laura revealed all to her, confirming that her father was indeed half Skanicum." Mary

Louise supposed to look relatively normal, but to have a wide mouth, protruding teeth, and "squint" eyes.

- o Daughter Madeline [Christine], living on the "Washington coast" in 1985, was considered ugly by Indian standards; shared the wide mouth, protruding teeth, and "squint" eyes of Mary Louise; but also had the typical Skanicum sloping forehead and long peaked ears. Was said to spend too much time on the pursuit of alcohol.
- o Stella (most recent birthdate daughter), who also died young

- > Patrick's wife was Laura's first cousin
- > Francis (born about 1925) met Patrick when she was about eight years old (so about 1933)
- > Louie of the Moses-Columbia band, was 79 years old in 1985 (so born about 1906), knew Patrick and his family all of his life. Louie worked on Patrick's ranch about 1925-1930.

The list above is the total of what Moser and I had to go on as we began our research. We felt that there were enough hard facts in Dr. Fusch's paper that we had a shot at tracking Patrick down even one hundred thirty years later. This book is about that journey to find Patrick.

Before Patty

CHAPTER THREE

Dr. Ed Fusch

In a book filled with characters, Edgar Fusch is able to hold his own. Ed is our twentieth century link to the nineteenth century Patrick, so a closer look at Dr. Fusch's biography provides context for this incredible story.

Just as Heather Moser found in her initial research, I had trouble tracking down Ed Fusch and his personal history. He did not leave much trace on the internet, particularly considering what a colorful character he was in life. His presumed birth record as "Edgor Fusch" (first name likely a transcription error) has him born on November 8, 1932 in Billings, Montana to Frank and Minnie Fusch.[9] Jean-Paul Debenat, PhD in his 2009 book, *Sasquatch/Bigfoot and the Mystery of the Wild Man: Cryptozoology and Mythology in the Pacific Northwest*, described Ed as "approaching 60" when he met him in 1990.[10] Ed was actually 67 in 1990.

The 1940 U.S. Federal Census gives Ed an older brother and a younger sister and brother. In 1947 he surfaces again in a yearbook photo from the old Kent-Meridian High School (go Cavaliers!). Kent is a town between Seattle and Tacoma,

---

[9] 11/08/1932, Edgor Fusch; Montana Birth Record

[10] Debenat 2009, page 85

Washington in the soggy western half of that state. In those days Kent was a growing bedroom community with a bustling derivative economy dependent on the manufacture of Boeing airplanes. The 1950 census confirms that Ed's father was a carpenter.

An article about the Kent-Meridian baseball team places Ed as pitcher. The pictures I have of Ed from 1947 and 1950 show a slightly nerdy-looking young man without the glasses he would wear later in life. Temper my rash use of the word "nerdy" with the fact that this Ed played baseball all four years of high school. Return to the nerd suspicion when you read that he was also in the Latin Club in his second and third years at that school.

The old Kent-Meridian High School and the Cavaliers, were converted to a Junior High right after Ed Fusch graduated.

Ed was enrolled in college classes sporadically throughout his adult life. In an *About the Author*[11] (presumably written by Ed himself), "Enrolling as a Pre-Med student at the University of Washington in 1951, 'Prospector Ed's' education was soon interrupted by the Korean War." The implication being that he was actually *in* the Korean war. Not exactly, as you'll soon read.

Ed first married in June of 1952 when he was only 19. His bride, Martha Eunice Barton, was 15 years old. Another female Barton, perhaps Martha's mother, witnessed the event.[12] I know what you are thinking, but I find no record of a child born soon after to Ed and Martha.

Unfortunately, I did not find Ed in any census records after 1950. Census records include all family members living in the same household and so might tell us if there was an early child in

---

[11] Fusch 1992b, *Testing & Assaying Gold, Silver & Platinum book by Prospector Ed*

[12] 6/20/1952, Edgar Fusch & Martha Eunice Barton; Marriage Certificate; State of Washington

Ed's life. Somehow Ed managed to avoid the census takers throughout his whole adult life, as far as my research reached. To me, this is confirmation of the deep rebel streak in Ed.

Perhaps the pressures of married life, or the very real possibility of being drafted into the Korean war (fought June 1950 until July 1953), are reflected in Ed joining the Coast Guard in December 1952.[13] Ed would be discharged from the Coast Guard two years later, in December 1954, but not before a man of his name and age is booked into jail in Clallam County Washington (a county on the straights of Juan de Fuca with Coast Guard facilities) on 7/30/1954.[14] Was the future doctor involved in some minor altercation? Perhaps a bar fight? The record does not clarify, but the booking probably did not help his Coast Guard career and a few months later he had served his two years and was discharged, still a youthful 22 years old.

The next promising lead is of someone by the name of Ed Fusch identified in a newspaper article as an usher at a June 1955 wedding in Spokane, Washington (clear over in the far eastern part of the state).[15] The Ed I am tracking would have still been 22 at that wedding.

By 1959 it seems Ed was getting closer to the American Dream. He is listed in the Seattle City Directory of that year as still married to Martha and manager of Modern Appliance Sales and Service.[16] In 1960, Debra Kay Fusch was born to Ed and Martha in June, when Ed was 27 and Martha 23. I found no record for other children born to the couple for another twelve or thirteen years.

From the same 1992 *About the Author*: "For the next 35 years after the war [the Korean war ended in 1953] he operated small businesses in the Seattle area and raised a family."[17] That bland summation may be technically correct, but glosses over some

---

[13] There was a mandatory military draft from June 1950 (when Ed was still only 17) until July 1973.

[14] 7/30/1954, Edgar Fusch; Register of Prisoners Confined; Clallam County; Washington State Corrections and Jail Records

[15] 6/25/1955 Ed Fusch; *The Spokesman-Review*, page 3

[16] 1959, Edgar Fusch & Martha E; City Directory; Seattle, Washington

[17] Fusch 1992b, *Testing & Assaying Gold, Silver & Platinum…*

harsh life changes in those 35 years between 1953 and 1988. There is a death certificate granted in November 1973—still in King County—for a new baby to the couple, Ann P. Fusch, just days after Ed's 41st birthday. Martha would have been 36 or 37 years old.

Marriages sometimes do not survive the death of a child and, less than three years later on March 20, 1976, Ed married Marlene A. Warner in Grays Harbor County, though the marriage license was granted in King County. I was not able to find a Certificate of Dissolution for Ed's first marriage to Martha Eunice Fusch. The second marriage to Marlene was not a keeper and divorce was granted to the couple just short of seven years later in 1983.[18]

By 1985, Ed was interviewing residents of the Colville Reservation about Skanicum and the abduction that led to Patrick. It would be some years before he would publish his findings.

Outright tragedy strikes when the first and remaining child of Ed's first marriage, Debra Kay, dies in extremely troubling circumstances on December 28, 1990, about 17 years after Martha and Ed's second child (Ann P.) had died in her first year. First child, Debra Kay, died at her own home when only 30 years old, apparently already in her second marriage. She was buried near the old family residence in Kent, Washington. Ed was then 58 and Martha 53 or 54.

According to the Certificate of Death[19], Debra Kay's mother, Martha Eunice, was now back to using her maiden name of Barton.

Earlier in 1990—before the death of Debra Kay—Ed had met Dr. Jean-Paul Debenat, a French professor of comparative literature. Debenat would later publish a book on Sasquatch in 2009. Fusch appeared prominently in that work. Ed was described in the translator's note as having "inventive idiosyncrasies."[20] In the beginning of the four Debenat interviews starting 1990, Ed was well into his anthropological research on the Colville

---

[18] 2/07/1983, Edgar Fusch & Marlene Ann Fusch; Certificate of Dissolution or Declaration of Invalidity of Marriage; Vital Records; State of Washington

[19] 12/28/1990, Debra Kay [redacted] aka Debra [redacted]; Certificate of Death; Vital Records; State of Washington Department of Health

[20] Debenat 2009, page 9

Reservation.

Debenat describes Ed at this time as having a commanding presence, a loud voice, "sturdily built," gregarious, and quite happy to dominate the conversation at a pub gathering Debenat attended. Debenat wrote that Ed was "a lively and gifted teacher, Ed keeps his audience enthralled."[21]

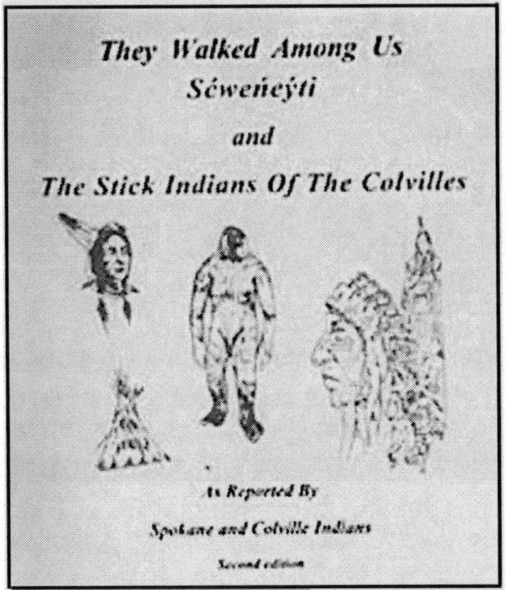

They Walked Among Us: Sćweńeýti and the Stick Indians of the Colvilles; *As Reported by Spokane and Colville Indians*; second edition. –The cover of Ed Fusch's 1992 spiral bound paper, cropped and enhanced for clarity

In 1992 Ed published his small "book"—cover pictured here (actually a medium-length paper enhanced with maps and illustrations). It was this paper that broke the Patrick story. In some versions it is titled, *S'cwene'yti and the Stick Indians of the Colvilles: The Interaction of Large Bipedal Hominids with American Indians as Reported to Dr. Ed Fusch, Anthropologist.*[22]

Also in 1992, Ed wrote that he had, "simultaneously received four degrees in Arts and Science, Geology, Anthropology, and Philosophy." I'm not certain that Dr. Fusch ever actually earned a doctorate, but he certainly put in his time at higher institutions of learning and also did archaeological, geological, and vulcanological field research in Africa, Washington state, Kentucky, and Hawaii.[23]

Debenat first met Ed at the July 1990 conference "Fabulous

[21] Ibid., page 314
[22] Fusch 1992a
[23] Fusch 1992b

Beasts: Fact and Folklore" in Guildford, England where Ed presented his paper, *The Great Hominid Bipeds Described by the Colville and Spokane Indians*.[24] A short excerpt from the paper read, "...large bipedal hominids existed, lived among, and interacted with the Indians from ancient days right up to very recent times."[25]

By now, Fusch was living in Riverside, Washington, just outside the northwest corner of the Colville Reservation. Debenat was to visit Ed here in 1990, February 1991 (very soon after the death of Ed's daughter, Debra Kay), February 1992, and in June of 1994 as research for Debenat's book. In a follow-up meeting in July 2007, Debenat gave Ed a copy of that book. It is from records of these Debenat visits that I found the best sources describing Ed Fusch's unique nature and distinct personality during the time of his Colville research.

The inventor seemingly found no old equipment worthy of consignment to the scrap heap. Debenat described Ed as driving an old 4x4 6-cylinder pickup with "no muffler." The fuel pump would fail now and then and Ed would just push a button to start an extra fuel pump he had installed and the motor would roar back to life. Debenat wrote in French of Ed's typical running commentary during trips in this truck [translated], "...Ed's harangue, delivered loudly enough to cover the sound of the motor."

Ed's Riverside housing was no less individual. Debenat wrote that as of 1991—for ten years already—Fusch's home in Riverside was a former primary school that he was gradually improving. He used one classroom for his living room, kitchen, bedroom, guest room, and a sitting room—furnished with car seats! In a second floor classroom, Ed had a tomato hydroponics growing operation of 80 plants (in *February* and before hydroponics was cool). Another classroom had become his

---

[24] Also titled "Large Bipedal Hominids as Reported by Spokane and Colville Indians."
https://books.google.com/books?id=cG4bAQAAIAAJ&focus=searchwithinvolume&q=colville

[25] International Society of Cryptozoology newsletter volume 9, number 3, Autumn 1990

laboratory, where Ed analyzed minerals and precious metals as part of his diverse livelihood.

The school auditorium (Debenat described it as, "this cave of Ali Baba") had its use as well. Despite a leaking roof, Ed had converted the auditorium to a "store" with partitioned stalls under the balcony devoted to different types of goods such as books on geology, prospecting, and mineral assaying; metal detectors; and diving gear. In the middle of the hall were bulkier items such as dredges, pumps, conveyor belts, and aspirators.

Debenat described some of Ed's lifestyle and livelihood:

> ...in this month of August, my friend Ed Fusch frequently leaves home in Riverside to exhibit and sell his gold prospecting equipment at fairs, trade shows and flea markets. Summer is orchard and gold-digging season.[26]

On Ed's 1990 business card, he describes himself as an anthropologist, geologist, philosopher, and scientist. In case there was any doubt of a soft side to Ed, he kept multiple birdfeeders on the old school grounds stocked with sunflower seeds for the hungry pine grosbeaks.

By June of 1994, Ed had billboards on highway 97 advertising his gold prospecting equipment. In the same visit Ed fed Debenat with a homecooked meal of fried onions and chicken.[27]

Debenat clearly found Ed's truck a precarious means of transportation—at 60 mph, the wheels and

[26] Debenat 2009, page 136

[27] Ibid., page 314

transmission would shake alarmingly. It is surprising then to find that in February of 1992 (during Debenat's four-day second visit to Washington) Debenat himself was driving the old truck over the 3,252' (990 m) Disautel Pass on the Colville Reservation when a blizzard hit them. Debenat urged Ed to take the wheel instead, but Ed said, "Gee Whiz! We *can't* stop! We'd never be able to start again... The important thing is to stay on the road."

When they arrived safely at Ed's schoolhouse home, Debenat had a shot of bourbon.

The heat for the classroom living space came from a woodstove that Ed had made from a water heater tank. "Some rather debatable projects never came to fruition, probably for the best." But Debenat continued with the praise, "…there seemed to be no tool that Ed couldn't take advantage of, no technical difficulty would deter him, no obstacle ever set him back."

Ed conducted research with Native Americans on both the nearby Colville and on the further east Spokane Reservation. He wrote that the challenges of this research included the old age of his informants—with their poor hearing—and the uneven English speaking-skills of these informants. Despite the challenges, by 1990 Ed had already conducted seven years of field work on the reservations.

After Debenat's visits, Ed Fusch's trail grows faint despite his revelatory work uncovering the remarkable Patrick story. In a 2005 article in Colville's *Tribal Tribune*, "Mr. Fusch" is lauded for his "researched and documented stories and accountings of the Sasquatch as told to him by Elders of both the Colville and Spokane tribes. He has worked as an Anthropologist for the Colvilles and Spokanes some fifteen years ago."[28] Ed was 72 when this article was published.

Rhettman A. Mullis Jr. (associated with Bigfootology[29]) was able to reach Ed by email in July 2012 after months of trying. He immediately traveled to Omak/Riverside for a videoed interview with Ed in which they "discussed Patrick, his offspring…" I have not been able to find this interview.

---

[28] 11/02/2005, page 1; *Tribal Tribune*

[29] https://bigfootology.yolasite.com/

Soon after that interview Ed Fusch moved to Bouse, Arizona where he died on January 24, 2015 at the age of 82.[30]

[30] 1/24/2015, Edgar Fusch; U.S. Department of Veterans Affairs BIRLS Death File

Before Patty

CHAPTER FOUR

Escape From Skanicum

"These women were never happy when they came back to their people, as after a time they would long to go back to their devil husbands and their children. They always managed to get away and return to the old wild life, as it held such a fascination for them, when they once experienced the wilds, that they could not resist the calling of such a life."

---Lucy Thompson 1991
(original ed. 1916) page 130

Hahissiat Kolockun had lost track of the days, but she could tell by the reddish tinge on the edges of the huckleberry bush leaves and by the mushrooms emerging after the last rain that it was late summer. At least two moons then, since the Skanicum had taken her. Already bony, according to a teasing Pierre Paul, her shoulders, ribs, and hips gave her even less padding now when she tried to sleep on the collected ferns of the cave bedding. Sleep did not come easy being held captive by a Skanicum even with the hide blankets, stolen from the valley Indians.

She had always been taught that the Forest People were another type of human and deserved as much respect as any neighboring tribe, but this was too much: To be taken from her new husband, like a coveted cow hide, and hidden in the dark of a cave. Once in a great while, she was allowed to linger outside of the entrance for half an hour or two, hoping to hear the happy, far-off sounds of her people in the valley below. She chafed at this life, knowing that the cold and the dank darkness of the cave would kill her in the end. Hahissiat only had the clothes that she had been wearing when she went to fetch water at Fish Camp. The long dress was torn and stained now and she had a cough that would not go away. She did not see how she would live through the coming winter.

There was an old fire ring in the cave. It could have been made by her hairless people, but she suspected others of *his* kind could make fire and had done so in this very high place concealed in a hard-to-reach rockfall. She gradually came to understand that despite his giant size, this male was young—perhaps younger in his body than she was in hers—and he was too clumsy or too lazy to find a mate from his own kind. He might not even know how to make fire.

At least now he trusted her depth of sullen defeat enough to allow her outside, supervised, to help gather the tule and cattail roots that he was storing for the winter in the back of the cave. And recently he would allow himself to nap for a bit while she did the work of a slave, gathering wild potato roots[31], which just gave her a stomach ache if she ate too many, but the hairy one could eat huge quantities of the little round tubers.

She continued to work with her improvised digging stick around the edge of this marshy area, looking for the arrowhead shaped leaves. She would pry loose a potato, shake off the mud as best she could, and throw it in the growing pile. The Skanicum had eaten his fill of huckleberries in a patch upslope and then settled down for a nap in the bushes, leaving her to dig. It was not the first time. She knew that if she stopped digging long enough that the absence of sound would wake the great beast and he would check on her. So she kept digging.

She was not as beaten as the Hairy One thought—she was always looking for a chance to slip away. Once she had tried to

[31] *Sagittaria latifolia*

escape while he was out hunting deer, but the Skanicum was so fast, and his nose for Indian so good, that he had caught up to her long before she could reach her own people. He hadn't even gotten angry. It was like a game to him and he had won handily. She did not want to give him *that* satisfaction again.

One of the sacred dark-blue birds, that seemed always to give her company and hope when she needed it most, perched on a nearby fir branch and then winged off through the trees, across slope. She followed its carefree path with her eyes until she happened to notice faint movement in the distance. Peering closer, she saw several Indian men on horseback through the gaps in the open woods. They were a hundred yards away across the narrow mountain valley. By their careful dress and the horse tack, she guessed that they might even belong to Pierre Paul's people, the Moses-Columbia.

There was no time to waste—she quietly lowered the digging stick to the ground and began to sprint towards the horses. Hahissiat could not believe it could be this easy—that this was now the moment that she had prayed for all these weeks. She asked the blue bird to give her wings to carry her to freedom.

One of the horsemen darted his pony out towards her as she ran. With a start, she saw that this was the same man who had lowered his rifle at the creek that horrible day. All the better then if he should be the one that delivered her from this horrible trial. His rifle sheathed in a scabbard behind him, he turned his horse in her direction and leaned a forearm down low for her to lock her skinny arm onto. Using her own momentum to swing up behind him, she dropped onto the horse's broad rump.

Hahissiat leaned forward to whisper in his ear as he kicked the quarter horse into a trot, "Hurry! He is asleep, but will wake soon!"

The other riders did not hesitate as they fanned out to protect the rear and they raced through the trees for the lower elevations.

They did not talk the whole way down the mountainside, except to call out quick thoughts on the route out of the area, and to tell Hahissiat that they were headed for a camp below. There was a sense among the whole party that this was a dangerous flight. Skanicum was known to be able to run as fast as any horse.

They were all desperate to get as far from the sleeping giant as they could. They knew that their trail of churned ground would be easy to follow.

Picture from about the time of Hahissiat's abduction: Many women on the Colville were accomplished riders.

At the camp in a glade by a mountain creek, a mile above the bottom of the river valley, the men shouted orders to the few startled women, children, and elders. They were to pack up and leave immediately. The questions stopped when they saw the half wild and very dirty Hahissiat. Seeing her, their eyes widened with fear and they quickly began tearing down tents and packing bags.

Before the four men had loaded their rifles and took positions around camp, her rescuer explained, "I could not shoot to stop the Skanicum that day. They will kill twelve for every one of them that we hurt and you would have been the first one that he killed. I have looked for you ever since. But now we must get out of here as quickly as possible and get back to the village downstream where there are so many of us that the Skanicum won't dare to strike. Until we leave, stay out of sight. Get my wife," he gestured, "to give you some new clothes and put a shawl around your wild hair. We'll soon get you back to Pierre Paul."

She was beginning to believe that she really was going to escape from the Skanicum and the nightmare of the last two of months. But she wondered if she might be bringing something of the Skanicum along with her.

CHAPTER FIVE

The Colville Reservation

The north central Washington state setting for Hahissiat Kolockun's harrowing abduction, was the Colville Reservation. It's climate is not that of the seemingly ever-dripping western Washington, but instead that of the arid and semi-arid hill country east of the Cascade divide. Vegetation on the Colville varies from mixed evergreen and deciduous forests in the uplands to sagebrush and grasses on the southern slopes and hotter lowlands. Even at lower elevations, there are scattered stands of pines on northern slopes and in the drainages. Forest land covers approximately two thirds of the reservation. Elevations range from 790 to 6,774 feet (240-2,060 m).[32] To this day, irrigated and dry land farming are significant land uses on the Reservation, along with livestock grazing, forestry, and mining.

Today this 1.4 million acre reservation has about 8,000 residents (less people per square mile than Wyoming; more than Alaska, my home state). In 1892 the Colville had just over 3,000 native occupants. That number would have been much higher without the disastrous smallpox epidemics of the 1770s and 1820s racing before the white expansion and the follow-up efforts to subdue the twelve tribes in the mid-1800s.

[32] Fusch 1992a, page 17

Temperatures on the Colville range from average winter lows in December and January in the teens to thirties (Fahrenheit) and summer highs in July and August in the eighties and nineties. It is a land that once was rich with anadromous fish, journeying from the Pacific and up the huge Columbia river and its many tributaries such as the Sanpoil. But then came the introduction of fish canning in 1866—which was impacting fish harvests for Hahissiat and her peoples already in 1891—and then dams in the twentieth century eventually slowed the flow of fish to a trickle, eliminating this critical nutritional and cultural resource for the Native peoples of this enormous drainage. Waters in parts of Idaho, Washington, Oregon, Montana, Wyoming, Nevada, and Utah—and then across the international border into much of the Canadian Province of British Columbia—all drain into the Columbia river and then into the Pacific Ocean. In 1940, the reservoir of the Grand Coulee Dam began to fill and eventually flooded 18,000 acres of Colville Reservation land, including homes, ancient fishing sites, and cemeteries.

Decades before the dams, the Colville Reservation was ultimately formed in 1872 to pacify and compensate twelve tribes that had once ranged over nearly forty million acres of eastern Washington, northeastern Oregon, western Idaho, and southeast British Columbia. The Chelan, the Chief Joseph Band of the Nez Percé, the Colville, the Entiat, the Lakes tribe, the Methow, the Moses-Columbia, the Nespelem, the Okanogan, the Palus, the Sanpoil, and the Wenatchi all shared this large reservation.

In 1880, Chief Moses had moved his band of Columbias to the Nespelem valley while he continued to negotiate for his own larger "Moses Reservation," representing his tribe. Moses was also recognized as the negotiating chief of the Chelans, Entiats, Methows, Okanogans, and Wenatchis. In 1883 he traveled to Washington D.C. to cement the negotiations. After initially being told that he would get his reservation, in 1884 Chief Moses' band of the Columbias and his allied tribes had to cede the Moses Reservation and remain permanently on the Colville.[33]

---

[33] Kershner 2008. Chief Moses (1829-1899). https://www.historylink.org/file/8870; accessed 2025.10.01

Chief Joseph of the Nez Percé

Primarily remembered now for their previous dependence on salmon, some of these tribes would travel far east, beyond the Rockies, and hunt buffalo. The Nez Percé were renowned for their horses and, only fourteen years before Hahissiat's abduction, gave the U.S. Army a black eye and bloody lip during a four-month, 1,170 mile running battle in 1877—while protecting their vulnerable women and children. They were finally trapped just south of the Canadian border.[34]

In 1885 Chief Moses invited Chief Joseph and his outcast Nez Percé to move to the Nespelem valley, increasing his own prestige with his generosity. The two chiefs soon became competitive friends. Unfortunately, measles outbreaks greeted the newcomers.

On March 15, 1899, Chief Moses died at his home in Nespelem with all of his wives (including his favorite, Mary) at his side.[35] Inevitably, Chief Joseph also died in the Nespelem valley on September 21, 1904. Chiefs Joseph and Moses must have crossed the unusual child's path. It is interesting to wonder if they understood his origins.

In 1891, there were still polygamous marriages on the Colville Reservation.[36] War, disease, and alcohol had left several of the tribes short of suitable marriageable young men, which may explain the persistence of men with multiple wives. It appears though that Hahissiat Kolockun (later taking the English name,

---

[34] Note: Some of the Nez Percé, though already exhausted, escaped that final besieged encampment of Chief Joseph, and crossed into Canada and out of the grasp of the U.S. Army.
[35] Kershner 2008. Chief Moses (1829-1899). https://www.historylink.org/file/8870; accessed 2025.10.01.
[36] *Report of Colville Agency*; July 31, 1893

Madeline) never had to share her husband, despite the near unimaginable stresses that her abduction put on their marriage.

Before western culture began a shift in the Moses-Columbia habits, people used to rise at dawn to begin their day by bathing in the river. Their name for themselves: škwáxčənəx^{w} [author's phonetic attempt: SHQUAKT-j'nook], meant "people living on the bank."[37] The fate of the Moses-Columbia was completely tied then to that of the wild and bountiful Columbia river.

In 1891, the Moses-Columbia band of Hahissiat's husband were concentrated in the semi-arid Nespelem valley along with the Nez Percé survivors of "Chief Joeseph's War." The Moses-Columbia traditional territory had ranged over 4.3 million acres, but they adapted to the new reality of confined reservation life, raising grain, horses, and cattle, and growing some vegetables. According to the July 1892 report from the Physician at the Colville Agency, the Moses band were:

> …true, genuine Indians in every sense of the word. Still they are very easy to get along with, and very pleasant if kindly treated…. As a class I think these people are the most robust and healthy Indians on either reservation.

The proud Nez Percé, however—still reeling from their defeat by the United States Army—relied on government rations as they adjusted to their exile from their traditional lands far to the south.

---

[37]Colville History/Archaeology Department. Revised 2024. *Confederated Tribes of the Colville Reservation: A Brief History*; https://storymaps.arcgis.com/stories/bb31cd48d0284fa59d6f454cafabe962; accessed 2025.10.01.

CHAPTER SIX

The Culture of Skanicum According to the Colville Tribes

"...Indians have been greatly humiliated by the Seeahtlks' "vulgar sense of humor... [the Seeahtlks] play practical jokes upon them and steal their Indian women. Sometimes an Indian woman comes back. More often she does not, and it is even said by some northwestern Indians that they have a strain of Seeahtkl blood in them."

---According to Jorg Totsgi,
Clallam Indian and editor of *The Real American*:

We tend to think of Sasquatch as something new on the scene—bursting forth into public consciousness with that never convincingly debunked film of Patty in 1967. But there is, of course, a long tradition of Skanicum in many forms on this Turtle Island from thousands of years of occupation by Native Americans. In the Columbia river valley of the twelve tribes, ancient stone cobbles have been found carved with wrinkled foreheads, large flat teeth, flat noses, bulging eyes, and weak chins.[38] Bobbie Short in her posthumously published 2024 book, *The De Facto Sasquatch: Cultural Legacies and Military Encounters*, summarized the First Nations' traditions of Sasquatch as hair-covered, giant-sized, living in the mountains, nocturnal, whistling to communicate, stealing food, as having the ability to hypnotize the Hairless Ones, to change their form at will, and a tendency to kidnap people.[39]

In the interior of Washington state, some specific traditions from the pre-European times survive. Both the Coeur d'Alene and Spokane Indians made the interesting observation that Skanicum will occasionally wear buffalo skins. The Coeur d'Alene believed that there were both dwarves *and* giants living in the mountains, but that the giants, in particular, took pleasure in scaring the Indians, such as by rock throwing. The Okanogan accused the Skanicum of engaging in "harmful" behavior. Northwest tribes in general felt that Sasquatch avoided whites and disliked their smell.[40]

Many tribes note the ability of Sasquatch to mimic other animals or to transform into trees or bushes. And many also reported on their unpleasant odor ranging from burning gunpowder[41] to wet grizzly bear (in the Yukon).[42]

Of particular interest to those tribes whose lives were so dependent on salmon, both the Coeur d'Alene and the Spokane reported that Skanicum liked to eat fish.

Explaining Sasquatch requires also explaining how they make it through the lean months of winter. For the First Nations perspective on that mystery, Short quotes Canadian Chehalis Indian, Henry Napolean, as saying that they sleep like bears far from the white men,[43] presumably in a dug den or a convenient cave. Some northwest tribes even suggest they wear bear skins to help them make it through the cold months. Burtsez reports that the indigenous belief in the Vyatka region of Russia is that the

---

[38] Meldrum 2006, pages 73-74
[39] Short 2024, page 28
[40] Debenat 2009, pages 350-351
[41] Ibid., page 350
[42] Grossinger 2022, page 37
[43] Short 2024, page 72

hairy man "hibernates like a normal bear, just starts it a couple of weeks later."[44] Many, including contemporary accounts, report that winter is a hard time for Skanicum. According to Northern Cheyenne elder, Otis Frank, during winter the Stick Indian women and children would become scrawny and gaunt. During plentiful seasons the Northern Cheyenne would trade with the mountain giants, but in the winter they would no longer ask for equal value back, but would instead leave high energy gifts of fat and seed. The kind gifts would be acknowledged with small tokens of appreciation.[45]

Dr. Jeff Meldrum has pointed out that pioneering Sasquatch researcher and statistician, John Green, found very little hard evidence, despite popular perception, that cave living is a significant element in Sasquatch natural history: "...indications of the use of caves, or any form of shelter, are very rare."[46] However, in my abduction research, the taken-people often report, or are believed, to have been held in caves. I will write about the cave topic more in the next chapter. The Coeur d'Alene and Spokane Indians believed that the Skanicum does live in caves. Perhaps we are seeing opportunistic behavior. If conditions call for cave-living or opportunity allows, they may be quick to take up residence underground.

The coastal Quinault feel that Sasquatch is half human and half animal, but also a protective being.[47] Skanicum may deliberately allow its presence to be seen—an exposure that may have a profound impact on the spiritual development of the eyewitness. The Nez Percé felt it was inappropriate to talk about Skanicum. Talking about them might draw their unwelcome attention to you.

Skanicum or Sasquatch, the native peoples of the Pacific Northwest had a long history of sharing their territories with something large that walked on two feet before Hahissiat was abducted. Bigfoot was a physical being that left traces in their environment, in their stories, and in their world view.

---

[44] Burtsev 2015, pages 147-148
[45] Short 2024, page 43
[46] Meldrum 2006
[47] Debenat 2009, pages 149-160

CHAPTER SEVEN

Abductions

"They look like a family—old man, old lady and two young ones, a boy and a girl. The boy and the girl seem to be scared of me. The old lady did not seem too pleased about what the old man dragged home. But the old man was waving his arms and telling them all what he had in mind."

---Albert Ostman describing his 1924 abduction in BFRO Report #1091

When I began seriously looking into the Patrick story, I was under the impression that there were few abduction accounts in the literature and that abductions were only something that had happened once or twice to human women, but not to men (with the famous Albert Ostman exception). I was wrong about all of that.

The more that I dug, the more I came to realize that a significant percentage of abductions happen to men, that abductions are a world-wide phenomenon, and that more than one species of "relict hominin"[48] may be involved, even in North

[48] See page 66 for definitions of hominin and hominid.

America. "Relict" in an anthropological sense refers to a formerly widespread population now reduced in numbers to remnants in a geographically relatively restricted range or ranges.

This book is specifically about a Sasquatch/Human hybridization in the Pacific Northwest, so I am not going to get sidetracked into the confusing worldwide situation. I will save that confusion for Volume II, where I seek an evolutionarily plausible explanation for Bigfoot in the global setting.

I think we would all prefer not to believe that getting abducted by an eight-foot tall hair-covered "monster" is one of the many things we have to worry about. Men in particular are quick to discount this as a concern. We are at the top of the food chain after all. Right?

Maybe top of the food chain in your city park. *Maybe.* In a large National Park? Nope. One encounter with a moose, bison, or a bear—black or grizzly—will set you straight on that account. A lot of people go missing in our wooded and lonely areas. Our First Nations people were (and still are) keenly aware that there are some regions you just don't go into.

Ron Morehead wrote in his 2024, *Bigfoot Unveiled: Scientific Answers to Bigfoot Mysteries*: "...because I believe that many Native American legends started with a core of truth, I think that some Bigfoot beings have crossbred with indigenous people..."[49] I will grant, as Paulides points out, there is not much in the way of confirmed abductions or even abduction attempts by Skanicum from the last fifty years.[50] But there are parts of western Canada where young women are warned not to walk alone, especially at night—areas with a horrendous number of unexplained disappearances. I would prefer to believe Sasquatch has nothing to do with those horrible events, but I leave Sasquatch on the suspect list.

Kewaunee Lapseritis—thoroughly in the "Friendly Forest Giants" camp—writes, "The Sasquatch people and Ancient Ones [a supposed distinct branch of relict hominin that is said to be extremely advanced spiritually] have told me that occasionally

---

[49] Morehead 2024, page 24
[50] Paulides 2017, page 473

rogue members of their group have taken women from our society when there is a shortage in theirs. The beings insisted it was extremely rare and they do not approve of kidnapping."[51]

Rare or not, there are some common themes in the reports of human abduction: People alone are the most vulnerable; caves as the scene of the imprisonment; improving the Sasquatch gene pool; out of necessity to find a mate; acts of sexual violence; mainly a western North America possibility; usually mentally destabilizing for those humans that do escape; and some acts may be motivated solely by nutritional needs (most of the abducted may just be quickly *eaten*). There is a subset of reports from among native populations of pine pitch being used to seal the eyes of the abducted so that they will not be able to find their way home again or reveal the Sasquatch strongholds, should they escape.

There is another subset of abductions, particularly for children, that appear to have a requirement of consent. Christopher Noël tells of Denise from Nova Scotia, recalling a childhood encounter:

> I had this feeling like if I went with them I was never going home again. It was like a fork in the road. Do I stay with my human mum or do I get adopted by a family of Bigfoot? She [the adult female Sasquatch] realized how scared I was. …and they just allowed me to run away.[52]

Before we are quick to condemn the whole Skanicum race [probably more accurately—*races*] due to this phenomena, have you ever heard of a bad human, capable of abducting or hurting people? It happens. Even with this abduction awareness, I still feel much safer in the woods than I do in congested cities. And I hike alone through grizzly country often. Take any group of like objects and they are subject to a bell curve depicting the range of their typical qualities. No doubt the bell curve of Sasquatch mindsets reflects a concentration of the great mass of Skanicum as peaceful, if secretive, coexisting with humans. There will also be a very few on the far right side that actually want to see us

---

[51] Lapseritis 2011, page 12
[52] Noël 2019, page 55

happy and growing in personal power. However, at the other side of the bell, will be those stinkers that want nothing to do with us and, given a chance, are happy to see the occasional hiker/hunter/hitchhiker/fisherperson come to harm. From Thompson again:

> The Klamath Indians, in bringing down their legends from the creation of man until the present day, say that some [Stick Indians] were made to be good and honorable, some bad; and some were real bad and mean, which they termed devils, or Oh-ma-ha.[53]

The threat of abduction is considered very real in many parts of the West. Red Grossinger—the dedicated but recently deceased Yukon-based "Nahganne" researcher—reported the words of Grant Pauls from northern British Columbia:

> I heard many stories from Kaska elders in which they were described as giants… that stole women and children, and you had to be wary of their potential presence while hunting, fishing, or trapping in the bush.[54]

But also, far away in New Mexico, the Zuni saw the *Atahsaia* as a cannibal that abducts children to eat and women to eat or make wives out of. Captives are released now and then.[55] The nearby Navajo, according to Short again, had to contend with the *Ye'iitsoh* or *Yei'tso*, *formerly* a human-eater who would abduct child-bearing age women.[56]

That's the big picture of abductions by Sasquatch in North America: Formerly a very real and terrifying threat, particularly for the women natives living here; apparently very rare everywhere that European culture (and firearms?) have penetrated. I have some specific abduction accounts with some interesting clues to look at before we get back to Patrick. Let's

---

[53] Thompson 1991 (original ed. 1916), page 129
[54] Grossinger 2022, page 254
[55] Short 2024, page 100
[56] Ibid., pages 127-128

start with abductions of human males, taken chronologically to maintain the historical context.

"A couple hundred years ago" [Written in the late 1980s in the 1992 Fusch paper, so ~ **1780s**?] a Wenatchee man was near Lake Wenatchee when he was captured by *Choanitos* and held in a smelly cave through the winter. "They were like a different tribe of Indians." He was returned in the Spring to the spot where he had been captured. He had been well-treated.[57] This tendency of Native Americans to think of Skanicum as just another tribe, with special abilities, living a rough life, persists through many accounts. Later in this book, I will look at hunter-gatherer strategies for survival and compare those with Sasquatch strategies.

Paulides has a horrifying account from **1915** near Hyampom, California. A two-person team of railroad workers were making the initial brush and tree clearing of a new route in deep woods. At the end of the day, one man was missing. An initial search found nothing. However, four weeks later he was discovered naked and near death at the bottom of a hand-dug pit. The soles of his hands and feet had been abraded to the point of rawness, as if by the action of heavy grit sandpaper. The near lifeless man was able to express that he had been captured by a female Sasquatch, that she had licked his hands and feet raw to prevent his escape from the pit, and that she had forced him to have sex with her. Tragically, he died soon after his rescue.

We started off this chapter with an introduction to the **1924** Albert abduction when he was approximately thirty-one years old. Ostman's is a very solid account. He was interviewed over decades by several people. According to the pioneering researcher, John Green, Ostman was interviewed by an anthropologist, a magistrate, newspaper reporters, a veterinarian, and a zoologist. No was ever able to find a variation in the Ostman story.

Ostman was looking for mining prospects on the west coast of British Columbia, near Bella Coola, ignored the advice of his Indian guide, and put ashore in a place known for its Hairy People.

[57] Ibid., page 87

All went fine for several nights until, when sound asleep, he woke to the awareness that he had been picked up in his sleeping bag (as if in a huge sack), and carried for at least three hours. His pack was brought along too.

Ostman somehow kept a cool Scandinavian head—once he got circulation back in his limbs—and spent the next several days observing his captors (apparently a family unit of male, female, brother, and sister). He began crude communications with them. He had the impression that he was there to provide a mate for his captor's daughter.

Ostman also reported that they had woven bark-strip blankets, but had no signs of fire use.

His chance to escape came when he fed the adult male a possibly fatal dose of chewing tobacco. In the hullabaloo that followed, he was able to race out of camp, and a couple of days later reached people further inland.

There are three significant takeaways from this Ostman story: This group had a spoken language; the big fellow was having trouble finding a mate for his daughter; and these Sasquatch were capable of making fairly sophisticated products out of natural local materials.

The Ostman story—a rare abduction of a European human male—is one of the foundational stories of the North American Bigfoot world. You either accept it and Ostman's account, or you have to reason that his consistent repetition of his story as early as the 1950s (before Patty got Sasquatch rolling in the Anglo world) was a careful deceit or exaggeration. I am comfortable accepting his account verbatim with the possible exception of the start of his unique adventure at the end of the Toba Inlet. Looking at details of the account from Ostman, and making a site visit themselves, Morehead and others feel that Ostman was more likely dropped off on the lower Jervis Inlet.[58]

Only four years later, in **1928**, Short reports that an Indian man on Vancouver Island near the Conuma river was kidnapped and kept prisoner.[59] Apparently he did make his way back to his

---

[58] Morehead 2017 (second edition), pages 224-228

[59] Ibid., page 76

people.

Lapseritis had a conversation with a man, "elderly" in 1998, who claimed to have been abducted or willingly gone with the Sasquatch people around **1938**. This man had returned safely to the *Homo sapiens* world after some unspecified period of time. The informant refused to say exactly what had happened while he was gone.[60]

Sometime before or within the 1950s, Charley Victor of British Columbia, Canada—while out hunting—accidentally shot and incapacitated what looked to him to be a 12 to 14-year-old nude white boy. This encounter occurred near the town of Hatzic. He had thought that the noises coming from above him in a tree were some type of small game. Victor shot the instant that he flushed the noise-maker out of an overhead cavity in the tree. The boy did not use intelligible speech, but yelled out into the woods with what seemed to be a call for help. From the far distance came a booming response. Within half an hour a 6' tall, "almost Negro black" female Sasquatch came out of the woods and took the still living boy away.

This account would appear to be evidence of an infant Caucasian boy abducted from his parents before he could speak, pointing to a disappearance in the **1930s or `40s**. The Sasquatch female was short enough that she may only have been an adopted big sister or possibly an intended mate.

The validity of this next report is uncertain. It stuck in my head from first hearing of it in the 1970s or `80s. Perhaps in the **1960s**, a male college student was taken captive by a family of Sasquatch in southern British Columbia. You may find a brief mention of this account online; take it with a grain of salt. A couple of the details are so singular, I do not get the sense that they were made up to improve believability of the account.

As the story goes, the young man was abducted and found himself with a family group, the mother becoming his primary protector and benefactor. Unlike Ostman, he did not sense that he was breeding stock, but rather said that he was treated as "the beloved family *pet*."

---

[60] Lapseritis 1998, page 97

The story grows even more remarkable: The family took their pet with them to some sort of Sasquatch gathering. He found himself in a cave system, apparently fronting on an outdoor area filled with many Sasquatch, some so terrifying to the young man that he literally leapt into the arms of the mother Sasquatch for comfort.

Interestingly, the college student reported that the stinking Sasquatch caves were illuminated by fire! This is the most recent account that I have seen of fire use.

Somehow the kid got loose and began a rapid flight downhill to get back to the human world (some 20-30 miles, as I recall). Downhill is generally the direction of the human world and uphill leads to the Skanicum world.

As the story goes, he did not want to talk about this encounter at all, ever, once the initial story got out.

The next account comes from Mongolia via Igor Burtsev: A Mongolian teacher was abducted in the **1960s** and held in a cave for two weeks by two female **Almas** (Russian local name). He escaped when they lost interest in him.[61]

I find this next, tenuous, anecdote fascinating because the scene is only fifty miles from where I live now and I have set eyes on the location many times over a span of forty years. I worked with Jessie Desmond on organizing and holding both of the Boreal Bigfoot Expos in Fairbanks, Alaska. She is a researcher and author in her own right with a couple of books to her name. Jessie was in the virtual room when Heather Moser made the initial Patrick presentation to our group. Jessie focuses on cryptids and UFOs of the North. As part of her digging for stories, she spent some of her time hanging out in the Nenana Visitor Center, hoping to leverage her one-quarter Alaska Native ancestry into some candid conversations on local lore with the local Athabascans. She did get a couple of very interesting stories for her trouble that were set in this small town of Nenana, Alaska. One of those stories is almost endearing. The story I will share here is not endearing.

A resident male teen or young man was abducted sometime in the **1960s or `70s** in Nenana itself. Nenana, a river-shipping hub,

[61] Burtsev 2015, page 98

is set at the confluence of two rivers—the Nenana and the Tanana. As the story goes, one day, or night an adult female Sasquatch snatched up the young man right in town. Apparently the struggling young man made enough noise that some of the locals were able to see, but not prevent the abduction. The Sasquatch, abandoning cover in favor of a quick getaway, made the fateful decision to swim across the enormous and deep Tanana river (about a hundred yards across), for the relative safety of the wooded hills to the north. With the thrashing man held firmly against her body, she plunged into the greenish-gray and cold glacial melt waters and began her awkward crossing. Unfortunately, she neglected to keep the pinned man's head above water. He drowned in her grip! Eventually she realized that she was holding a corpse and she released him. His body drifted downstream with the current. Now unencumbered, she quickly finished the river crossing and escaped into the woods above the railroad tracks on the far shore.

The previous were nine accounts of boys or men being abducted by Sasquatch, in North America and in Asia. Three of the tales were from within my own lifetime. In seven of the stories, the men lived and were able to leave the wild world. In one case the man died during an attempted abduction. In another case, a presumably abducted modern human boy was injured when mistaken for a wild animal. It is unknown if he survived the injury.

I will use the anthropological term "modern human" frequently in this book. Modern is the word I use to distinguish us hairless *Homo sapiens* from these large and relatively hairy beings living in the woods. Sasquatch may also deserve to be considered *human*.

Of course, it is not surprising that our stories are those of the men that got away: Men that never get away don't get to tell their stories. Plenty of men (and women) go missing without a trace. For every man or woman that got away from Sasquatch, how many never escaped, except through death? Five? Ten? Twenty? I would prefer to be dealing with Friendly Forest Giants myself, but the facts show that we sometimes have something else going on. Something far more complex and unsettling.

Let's turn now to the accounts of abducted women. There are more of those reports and they are more relevant to our story about Hahissiat Kolockun and her child Patrick.

Paulides, again, is our source for the report of a woman **pre-1856** being abducted from the Fort Vancouver area (on the banks of the Columbia river, in southwest Washington state) by a *group* of *Tsiatkos* (Sasquatch). The woman later escaped.

In another abduction attempt, apparently from the same **pre-1856** period and location, an Indian shot and wounded a *Tsiatko* that was trying to carry off a "young girl."[62]

A 17-year-old Salish Indian woman at the Harrison river of British Columbia, was abducted in **1871** and held captive by her Bigfoot abductor for a year in a rock shelter [a deep overhang in a rock face] that they shared *with his parents*. She reported that she was not treated unkindly and was eventually returned to her home because she was too aggravating to her captor.[63]

Next is the often-cited case of Seraphine Leon or Long. Apparently about **1880,** Seraphine was kidnapped by a Sasquatch and held for a year in a cave. Heavily pregnant and now ill, she was finally returned to her home on the Douglas Reserve of southern British Columbia. Soon after, she gave birth to a baby that did not survive for very long.[64]

In **1891**, newly married Hahissiat Kolockun is abducted from a native fishing camp by a male Sasquatch on the Colville Reservation. She is kept in a cave for about two months before escaping, but returns pregnant.[65]

There is a report from **1899** of a 14 or 15-year-old Karuk Indian girl abducted by a Bigfoot in northern California. Thankfully, she comes out of the woods two weeks later, "dirty and tired."[66]

Meldrum had a report through J. Robert Alley (of *Raincoast Sasquatch* fame) of an elderly Coast Salish woman from Nanaimo, British Columbia (on Vancouver Island) that told him in 1974 that in about **1900**, her grandmother had been abducted

---

[62] Paulides 2017, page 429
[63] Alley 2003 (2018 edition), page 125
[64] Lapseritis 2011, pages 12-13 and Short 2024, pages 60-64
[65] Fusch 1992a, pages 19-22
[66] Paulides 2017, pages 431-432

by *Squee'-noos*. Her grandmother escaped, but gave birth to a malformed stillborn baby.

Meldrum also quotes another report by wildlife biologist, John Mioncyznski, among the eastern Shoshone: "a remarkably similar incident". Meldrum gives no further details, but I think it is safe to assume that there was a native woman abducted in the **early 1900s** somewhere in the Northern Rocky Mountains. The woman escaped, and there was a resulting non-viable pregnancy.[67]

Perhaps also from the early **1900s**, there is a detailed abduction story still told in the Russian village of Glinkovka (in the Kuzbass region) of a girl taken by a half man/half bear and confined with him in a cave. Her abductor would cover the entrance to the cave with rocks. She was held captive *for years* and gave birth to multiple children. Somehow, she escaped one day with all her hybrid kids.[68]

In the Foreword of Red Grossinger's book, Raymond Yakeleya tells the story of how in about **1930**, Marina Francis' daughter was taken from near Tulita, Northwest Territories, Canada and held for two or three days before escaping. Unfortunately, she died not long after her self-rescue.[69]

An anonymous writer's letter was read on Steve Isdahl's "How to Hunt" podcast of local knowledge in northern New Mexico. The story was that a woman was abducted in the **1930s**, but able to get back after some time, but she was now pregnant. She bore a son that came to be known locally as *Osito*—or "Little Bear"—due to his hairiness.

In the **1940s**, near Fredricktown, Missouri, a young girl disappeared, but returned two days later saying that she had been abducted by an "ape-man." According to the story, she became a nun.[70]

About **1950**, a Lillooet woman (not very far from Harrison Lake and the Douglas Reserve area of Serephine's experience) returns to her tribe heavily pregnant. She soon gives birth to a very

---

[67] Meldrum 2006, page 77

[68] Burtsev 2015, pages 382-383

[69] Grossinger 2022, pages 3-4

[70] Paulides 2017, page 435

hairy baby that did not live past the night that it was born.[71]

From Raymond Yakeleya again through Grossinger: About **1950**, his young aunt in the Northwest Territories of Canada was being carried away from a subsistence camp, but her father chased the Sasquatch until it dropped his daughter. Sadly, the daughter still died a week later.[72]

This next account is intriguing, but rests on the validity of some of the paranormal ("woo") associated with Sasquatch that I am going to keep at a distance until Volume III. However, this is such a poignant and relevant story, I can't resist including it.

As the story goes, in 1994, a human that was adept at communicating with Sasquatch through "mind-speak" (non-audible mental conversations) was in a forest clearing meeting with a male Sasquatch when a fully *modern human* woman walked out of the woods and joined them. Soon after a Sasquatch mother and infant swelled the confab. The modern human woman was having difficulty speaking verbally, so they soon were back to using mind-speak to communicate. This woman was in mis-matched clothing and was unkempt, her hair knotted and messy. During the conversation, the informant got the impression that this woman's name was "Sally" and that she had been taken by Sasquatch in northwestern Oregon when she was five or six years old. Even then, as a child, she was soon given the option of being returned to her own kind, but she had chosen to stay with the Sasquatch. The informant went on to encounter this singular woman on one other occasion.[73]

I have made initial attempts to track down missing persons reports or newspaper articles about a girl named Sally vanishing in the **1950s** somewhere in the area of the Oregon counties of Benton, Clatsop, Columbia, Lincoln, Polk, Tillamook, Washington, and Yamhill. No luck so far.

This next account has made the rounds because there is video support. Supposedly a human female lived through an abduction by multiple Yeren in Shennongjia, Sichuan Province, China about

---

[71] Ibid., page 430

[72] Grossinger 2022, page 2

[73] Lapseritis 1998, page 96

1960. The abduction left her pregnant. Video filmed in 1986 was found when another woman was going through the belongings of her deceased father. This video is apparently of the hybrid child at about age 23. The subject of the film was undeniably odd: Refusing to wear clothes; about 6'-5" tall; restless; reputably unable to speak, but able to understand; small brain case; and with unusually long arms. The subject is believed to have died about ten years later at the age of 33. Watch the video linked in the footnote or in the Bibliography and see what you think.[74] This video may be suitable for a limb proportion study. Perhaps ThinkerThunker has already done so.

This next story came to me through another acquaintance, Larry "Beans" Baxter. Beans is an Alaskan researcher, podcaster (Alasquatch), and author. In a note from Beans he recounts a story he heard that probably dated from the **1970s or `80s**. This story appears to be more of an oddly consenting interspecies-fraternization than of an abduction:

> When I first got to Alaska, and was stationed at Fort Rich [Fort Richardson]—before I met my wife—I was dating this Native girl. Somehow the subject turned to Bigfoot and the unexplained. I was probably telling her I liked X-Files or something. She tells me a story about a pilot and his wife who lived out in the bush. She [the pilot's wife] didn't like living in the bush and he loved it. But he was gone all the time because he was a pilot, and would always fly fisherman to and from places to fish.
>
> She started to go crazy, and he came home one day, and she was knitting a giant sweater—like a huge one—and he asked her what she was doing and she said she was knitting a sweater for her boyfriend. He kind of wrote it off thinking he needed to take her somewhere to get her checked-out, but he had some more work to do and a couple more charters to fly. I'm not sure on the timeframe maybe it was a day or two later maybe it was at the end of the season but he's flying home in his plane, and he sees his wife being carried away by a giant Hairy Man. If I

74 https://www.youtube.com/watch?v=OObRU_XMh14 and Redfern 2016, page 115

> remember correctly, there were three of them [Hairy Men] running away from the cabin. One was carrying his wife.[75]

As I was writing the first draft of this book, two very recent (**1984**) and oddly located accounts made the light of day for me as they showed up initially on Nexus News. The link in the footnote below has the full story (somewhat confusingly written in terms of numbers and chronology). A [teenage?] "girl" was reported missing along Cherry Creek in or just south of Denver, Colorado, but turned up again quickly. She explained that a Bigfoot had abducted her by carrying her over its shoulder. In 1984, Denver already had a population of over 500,000. Cherry Creek enters the Platte river in downtown Denver. I have biked the lower reaches of Cherry Creek and found nothing wilder along the concrete bike trail than homeless people and a muskrat. However, go far enough upstream and you do start to break out into open but mostly treeless country as you progress towards the large reservoir and Cherry Creek State Park to the southeast of Denver. The riparian corridor of the creek and the perimeter of the reservoir have bands of trees and brush where a young male Hairy One might be able to travel for his mischief. Mostly concealed at night, he might be mistaken for a large man by the occasional passerby.

The second case is of another missing teenage girl (also **1984**), but this time in the Wolhurst area of the South Platte river, about 20 miles south of downtown Denver and perhaps 15 miles west-southwest of the first short abduction. In this instance, a search was initiated, but had just begun when this girl again appeared, smelly hair still clinging to her jacket. From the Nexus article (attributed to someone that had joined the search party):

> A girl claimed that while she was out for a walk, something rustled through the bushes, grabbed her, and threw her over its shoulder. She showed us the jacket she was wearing, which was covered in a strange-smelling hair that was unlike typical fur. It was thicker than human hair and had a distinct odor that

---

[75] Baxter 2023, personal correspondence

> resembled wet bull… I even plucked a couple of the hairs from the jacket, and they had a horrible smell… the search and rescue team found her just as we arrived, and she was unharmed. [76]

Showing off her jacket to strangers does not sound like the response of someone deeply traumatized. This setting of suburban Denver is also odd, but maybe not out of the question. Both the South Platte and Cherry Creek are nestled in greenbelts. South of Denver, Wolhurst along the Platte is flanked by large reservoirs and is immediately north of the Chatfield Reservoir and State Park. From Chatfield, it is only a couple of miles to rugged wooded foothills that could have provided cover all the way to the continental divide of the Rockies. Traveling west less than twenty miles as the crow flies, this harry Lothario could have reached an elevation of 11,500'.

These two clumsy abduction attempts might have been by a young male Sasquatch (more commonly called Bigfoot in 1984), on his own and trying to start a family. His "seductions" were gawky and, fortunately, apparently ineffective. He might have been wandering the local creek trail systems as his way to "meet" potential partners.

These previous accounts from 1984 may have left traces in the *Denver Post* newspaper. My initial attempt did not find anything online. Research might also uncover other unexplained disappearances off these trails through the 1980s. It could be that this Bigfoot's relative gentleness grew more forceful and successful over time. Any resulting missing persons would have been blamed on modern human predators.

The two young women involved are probably still alive. It would be interesting to speak with them.

In the **Fall of 1988**, at Lake Sheldon in the Yukon Territory of Canada, "a large and tall bipedal entity" walked into a First Nation's hunting camp, picked up the hunter's daughter (she was about ten years old), and immediately began to walk off with this human theft. The hunter began shooting over the head of the

---

[76] https://phantomsandmonsters.com/post/1757167149323

Nahganne (a Yukon native name for Sasquatch). The *Nahganne* dropped the daughter and left, never to return.[77]

On October 13, 2004, the Russian paper, *Pravda*, published a long report (from the **1990s?**) about 19-year-old Oksana Terietskaya. For most of a year, she was held in a cave by the Russian version of what we would call a Sasquatch. The Bigfoot would carefully seal up the opening to the cave with a large boulder every time he left to forage or hunt. Gradually conditions and communications in her captivity improved for poor Oksana. But when her captor was careless with the boulder one day, she was able to squeeze out. She ran until she saw people.

Unfortunately, no one, even her parents, believed the young woman's story. She was deemed mentally incompetent and hospitalized in a sanitarium where she grew increasingly despondent and withdrawn. She had traded one prison for another. When even her doctor had given up on any hope for improvement, she suddenly brightened up and began to eat again.

Congratulated for her improvement, she had laughed and said that she had never been sick: She was just happy now because she knew that "he" was coming to rescue her. For this honesty, she was further isolated in a special room.

Early one morning of heavily falling snow, Oksana was missing from the sanitarium. The steel bars over a window opening had been pulled out of their brick anchors. Wardmates insisted that a giant hairy-monster had kidnapped the girl. The heavy snow made it impossible to follow tracks. Oksana has not been seen since.[78]

To recap this review of abductions, I located nine fairly solid accounts of *men* being abducted in North America and in Asia. In seven cases the men involved were relatively unharmed when they rejoined the human world. In one case (the man in the pit), the man was so injured, starved, exposed, and traumatized that he died soon after rescue. In the final case, a man died by drowning during an attempted abduction.

Men were abducted in the United States (four), Canada (four

---

[77] Grossinger 2022, pages 269-270

[78] Lapseritis 2011, pages 13-16

also), and Mongolia (one). Men were abducted in the U.S. states of Washington (two), California (one), and Alaska (one). All four men abducted in Canada were abducted in the province of British Columbia.

What I find fascinating and potentially highly significant, is that in at least four of the male abduction cases there is the potential of *human* genetic introgression *into* the Sasquatch gene pool. An introgression is when the genes of one species work their way into the genes of another species. Neanderthal and Denisovan genes have both been found in each other's populations and Neanderthal and Denisovan genes have both been found in us *Homo sapiens*. If a modern human male impregnates a female Sasquatch, and she bears a viable hybrid child that goes on to mate with another Sasquatch, and the products of that union go on to further gene-mixing in the Sasquatch world, pretty soon you have a Sasquatch population with human traces in their nuclear DNA (inherited through the father's side). After two or three generations the offspring may look like completely typical Sasquatch. Or such circumstances might explain the occasional bald or pale-skinned Sasquatch reported.

The cases of potential introgression *into* Sasquatch genetics include:

1) The poor man in the California pit in 1915. He told his rescuers that a creature (that we would immediately identify as a female Sasquatch) had imprisoned him for sex. There is a possibility that she became pregnant during the weeks of his captivity.
2) Lapseritis' had a reluctant informant say he lived among Sasquatch for a time about 1938, but refused to reveal how that time was spent. Perhaps he was coy because he couldn't bring himself to admit that he had impregnated a Patty lookalike?
3) The white boy in the tree, presumably abducted in the 1930s or `40s was apparently totally assimilated into Sasquatch culture. If he was able to recover from his gunshot wound, he may have wanted to mate in a few more years. I see this particular wild boy as a very strong

possibility of gene-mixing from modern human into a Sasquatch population.

4) The last possibility of an *H. sapiens* genetic introgression into relict hominins was the Mongolian man held by *two* female *Almas* (Asian term). That man said that he was able to escape when his captors lost interest in him. Interest in what? I think there was a good chance that they were interested in sex or procreation, or both. When they moved on in their cycles or became pregnant, they lost interest.

I have found a much bigger pool of abducted women than of men: 21 in all. Out of the total 30 abduction attempts, men experienced 30% and women 70%. For men, 22% of abduction attempts studied apparently led to their deaths. For women that figure was only 9.5% (two individuals, both in the Northwest Territories, both dying *after* returning from the abduction).

In the ten cases where we know where the abducted were held, eight or 80% were held in caves or rock shelters; one man was held in a pit; and another man was held out in the open, in a confined area with minimal points of egress, though the Sasquatch family themselves were using a rock shelter within this confined area (the Albert Ostman account).

Abduction attempts on women happened in the countries of the United States (11), Canada (7), Russia (2), and China (1). Abductions or abduction attempts on women happened in the U.S. states of Washington (3), Colorado (2), Alaska (1), California (1), Missouri (1), New Mexico (1), Oregon (1), and the northern Rockies (1). Abduction attempts happened in the Canadian regions of British Columbia (4), Northwest Territories (2), and Yukon (1).

Two abductions were foiled by rifle shots—either close to or wounding the Sasquatch. Of the remaining apparently successful abductions (19 in number), two of the women, or 9.5%, apparently never returned: 1) The middle-aged Sally reported living by choice with a Sasquatch clan in Oregon decades after her abduction as a child; and 2) Oksana Terietskaya, willingly returned to her captor in Russia after being neglected by the health care system. But 17 women or 81% *did* return. Of those 17 women, two soon died (9.5%).

Of the remaining 15 survivors, seven returned pregnant or had been pregnant during their captivity (47%). Of the seven pregnancies (I am counting the Russian case described as having "multiple" living hybrid children when she escaped as only one viable pregnancy), four had stillbirths or babies that died very soon after birth (57%). Three women had viable births that may have possibly gone on to have children of their own.

As far as an introgression into the *Homo sapiens* gene pool (by Sasquatch or some other relict hominin), these three viable births (in one case possibly "multiple" births) represent that possibility in recent times:

1) Sometime in the early 1900s, local knowledge accepts that a woman was held for years by a relict hominin creature in the Kuzbass region of Russia. Eventually she escaped with multiple hybrid children. If those children were strong enough to survive their early childhoods and escape to better conditions, it is likely that one or more of them eventually passed on their genes. 100 years later, there may be individuals out there with 1/8 and 1/16 blood of *something else*.
2) The half Sasquatch child, Osito, born in New Mexico in the 1930s, may have, despite his hairiness, found a partner and continued that particular introgression event. In that case, his quarter Sasquatch child or children may still be alive *today* and may in turn have 1/8 blood children living in our North American population.
3) Finally, we have the half-blood Patrick born in 1892 in Washington state. His birth was the start of a *significant* introgression event that will occupy much of the rest of this book.

Before Patty

CHAPTER EIGHT

The Birth of Patrick

> "During her stay with Skanicum the woman became pregnant and bore a son named Patrick, who grew up on the reservation."
>
> --Fusch 1992, page 20

I believe Patrick was conceived by about September 1891 during Hahissiat's abduction. By Thanksgiving, Hahissiat must have known that she was pregnant. She, and whomever she first confided in, did the math and calculated that she was going to bear a hybrid child around June 1892. This was an enormous predicament, not just for Madeline and her husband Pierre Paul, but for the entire family and even for the greater band of Moses-Columbia at Nespelem. The last thing they needed was for the steadily encroaching whites or their Indian Agent, Hal J. Cole, to hear that the Moses-Columbia had Stick Indian blood.

It's hard to imagine what this sixteen or seventeen-year-old woman went through in the weeks after she escaped. I believe her rushed flight out of the eastern Colville hills ended in Nespelem—already a substantial market town and the anchor for a small, mostly rural population of Pierre Paul's Moses-Columbia people and the banished Chief Joseph band of the Nez Percé. Nespelem is where Hahissiat, soon called by the English name, Madeline (with various spellings), spent the bulk of the rest of her life.

Other than Patrick's Delayed Birth Registration (issued ten years after his birth and reproduced at the end of this chapter), documents throughout his life declare Chelan, Washington as his birth place.[79] In the early 1890s roads on the Colville were few, in poor condition, and very difficult to navigate during rains or in winter snows. The Colville Indian Agent of 1894 wrote, "the maps showing the Colville and other reservations attached to this agency give the casual observer a very meager idea of the extent and scope of the country… There are no roads through this mountainous country…"[80] If a pregnant Madeline moved mid-winter, over one hundred miles by horse, carriage, steamship, and/or canoe (downstream on the Columbia river), she would have done so out of a great sense of necessity.

According to the 1909 Indian Census for the Colville Agency, both Madeline's widowed mother Mary Ann *Lakayuse* and her also widowed mother-in-law, *Nanuquiya* (also known as *Klaymeetk* and Catherine), were living in the same household with Madeline and Pierre Paul as late as 1909. These women would have both been about 60-years old in 1891. Presumably everyone was connected and in close communication, already knowing about the abduction and the unusual escape, and that there could be a very big problem headed their way.

Presumably, it was her mother, Mary Ann, that Madeline either spoke to first or that first realized that Madeline had a problem. Mary Ann might then have gone to her daughter's mother-in-law, *Nanuquiya,* and informed her of the pregnancy. *Nanuquiya* would

[79] 10/30/1912, Patrick [redacted]; Certificate of Birth—Delayed Birth Registration; Division of Vital Statistics; Washington State Department of Health

[80] University of Washington Digital Collection: Reports of Agents in Washington. Report of Colville Agency; August 21, 1894, page 314; Captain U.S. Army, Acting U.S. Indian Agent, Jno. W. Bubb

have then gone to her husband, *Squinimetsa*.[81][82] And finally the news would have been broken to her already suspecting new husband, Pierre Paul.

Obviously, something had to be done.

It was not even entertained by Hahissiat, her mother, or her mother-in-law, to use the services of a white doctor. Dr. E. H. Latham wrote in 1892, "The Indian women are very superstitious in regard to this branch of the profession, and would die before they would submit to a white doctor attending them in confinement."[83] So to depend on her mother's side of the family for help during her later pregnancy and labor would not have struck anyone as peculiar.

Moses-Columbia women circa 1940s: Sadie Paul Moses (1898-1977) and Lucy Moses Paul (1862-1948)

In the very old strategy of those avoiding scandal, Madeline had to leave Nespelem before it became obvious that she was pregnant She would have her child elsewhere and not return until a plausible story could be concocted.

We have already heard from Ed Fusch's Colville informants that Hahissiat Kolockun[84] was a "bride purchase"—presumably

---

[81] Another version of Patrick's step-grandfather's name is occasionally used as Patrick's last name. I think his step-grandfather's name, was more accurately Patrick's last name, however, the name appearing on most documents was a white clerk's effort that the family was ultimately stuck with in an Ellis Island type of phonetic misunderstanding.

[82] Second name from the 12/04/1933, Pierre Paul [redacted] (stepfather of Patrick), Certificate of Death; Bureau of Vital Statistics; Washington State Board of Health

[83] University of Washington Digital Collection: Reports of Agents in Washington. Report of Physician at Colville Agency; July 15, 1892

[84] Patrick's mother's name is given as "Madelein Nowkopop" on an October 1942, Patrick [redacted]; U.S., Social Security Application

from outside of the Moses-Columbia band. I tried to find document traces of *Hahissiat Kolockun* from before her abduction, but have not succeeded so far.[85] What I did find is her mother's name, Mary Ann *Lakayuse*, in the 1909 Indian Census.[86] It happens that there is a Lakayuse Road in Manson, Washington, less than ten miles from the town of Chelan. Based on this find, I suspect that *Hahissiat Kolockun* was of the Chelan tribe, one of the 12 confederated tribes of the Colville. A Chelan tribal affiliation for *Hahissiat* would make her journey to the town of Chelan, probably to her own mother's household, her best option.

To add a little uncertainty, I also found a woman living in Yakima, born about 1879 (compared to an estimated birth date for Hahissiat of about 1875) with a last name of Killockum—very possibly a spelling variation of Hahissiat's last name: Kolockun. On most Indian Census records, Nellie Killockum was full-blooded Indian (though on one census she is recorded as one quarter blood). Nellie was very likely a Yakama Indian. Possibly Hahissiat was also a Yakama Indian, even Nellie's older sister. And, as it turns out, Nespelem itself was founded by the great Yakama chief, Kamiakin (ca 1800-1877).

I believe it most likely that Hahissiat was a Chelan Indian, or perhaps a Yakama. Or both. Her father may have been a Yakama and her mother a Chelan.

In times of need, a young woman's natural inclination is to shelter with her mother or her mother's extended family. I believe her blood relatives were to the southwest of Nespelem in Chelan on the Columbia. Fortunately, Nespelem is near the Columbia river. Hahissiat probably made her way to and back from Chelan along that water route. Chelan relatives may have given her shelter for the last half of her pregnancy and for months after to confuse the age of the child.

Later in Patrick's life, he was to spend time even further southwest of Nespelem in Kittitas County (containing Ellensburg, Cle Elum, and Roslyn—home of the exterior scenes in the 1990s

---

[85] Scholars who want to try to find those traces may start with 1870s census records in Chelan and Yakima, Washington.

[86] Indian Census; Colville Agency, June 30, 1909.

television series "Northern Exposure") and in Wapato, just south of Yakima. In those southern towns he generally worked as an itinerant agricultural laborer. He may have gravitated to these areas because of the potential for support from his mother's side of the family.

Although Patrick consistently gave Chelan as his birthplace, he offered many conflicting ages and birthdates. From documents, his birthdate or year is given as—or may be calculated as—September 4, 1888; "about 1890"; September 1, 1892; September 1892; 1892; 1893; and 1898.

In actuality though, I believe Patrick was born somewhere around June of 1892, a little less than nine months after Hahissiat's escape. In the earliest document traces I find of Patrick, his English first and middle names are recorded as "Patrick Paul" (remember, his stepfather went by the first and middle names of Pierre Paul). Patrick also had an Indian name phonetically transcribed as *Wa-tam-i-ya-hle.*[87] This was the only time in many documents that I find an Indian name for Patrick.

It is likely that Patrick himself did not know his correct birthdate. I think that he, and the greater Nespelem community, were told that he was born in September of 1892. This date is long enough after his mother's escape at the end of the summer in 1891 to dampen rumors that he was a product of the abduction. But the date was also not so many months beyond his actual birth as to cause an obvious contradiction between his age appearance and his claimed date of birth.

The deception may have been aided by the fact that Patrick was very small. His reported adult height varied, but I believe he probably never grew taller than 5'-2" (157 cm).

Patrick was also odd-looking, presumably from birth. Ed Fusch reports his "body structure was different from other Indians… his arms were very long, reaching about to his knees… a sloping forehead… very large hands and long fingers…" In general, he was described as "very ugly." The poor hybrid kid was dealt a hard hand that was ultimately reflected in a hard life.

I found no good information to venture a solid return date, but

---

[87] Indian Census; Colville Agency, June 30, 1909.

eventually Patrick and his mother did return to Nespelem, perhaps around January 1, 1893, when Patrick was about six months old. Other than his birth in Chelan, Washington, I don't find Patrick anywhere but Nespelem until October 1928, when I believe he was 36-years-old. But he would spend many years after 1928 wandering around Washington state, leaving a trail for us to follow.

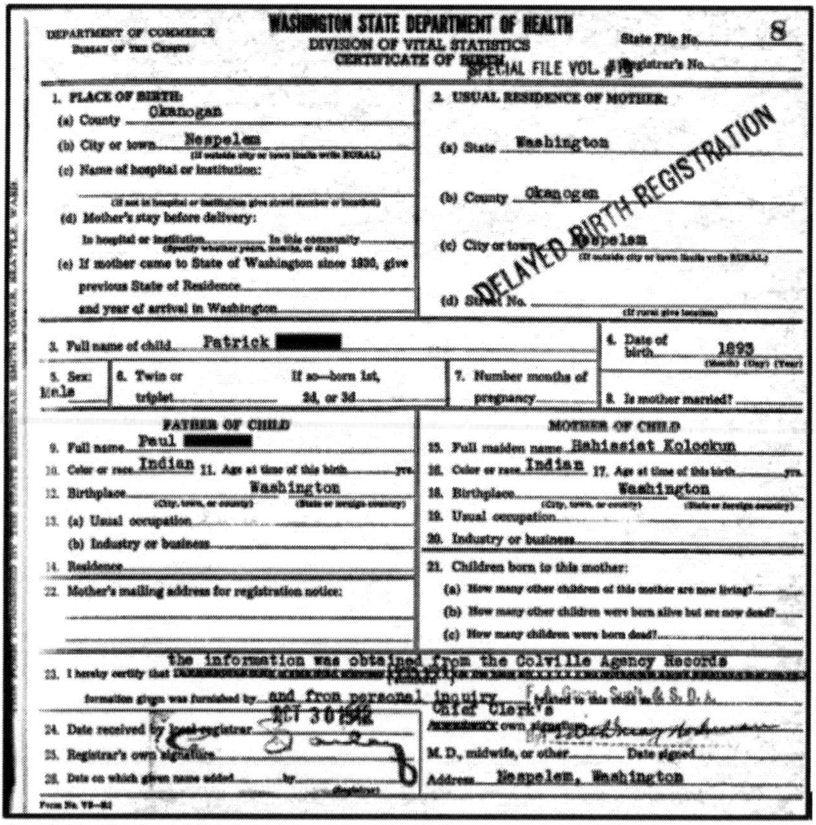
DEPARTMENT OF COMMERCE
BUREAU OF THE CENSUS
WASHINGTON STATE DEPARTMENT OF HEALTH
DIVISION OF VITAL STATISTICS
CERTIFICATE OF BIRTH
SPECIAL FILE VOL.
State File No. 8
Registrar's No.

DELAYED BIRTH REGISTRATION

1. PLACE OF BIRTH:
(a) County Okanogan
(b) City or town Nespelem
(c) Name of hospital or institution:
(d) Mother's stay before delivery: In hospital or institution ... In this community ...
(e) If mother came to State of Washington since 1930, give previous State of Residence ... and year of arrival in Washington ...

2. USUAL RESIDENCE OF MOTHER:
(a) State Washington
(b) County Okanogan
(c) City or town Nespelem
(d) Street No.

3. Full name of child Patrick [redacted]
4. Date of birth 1893 (Month) (Day) (Year)
5. Sex: Male
6. Twin or triplet ... If so—born 1st, 2d, or 3d ...
7. Number months of pregnancy ...
8. Is mother married? ...

FATHER OF CHILD
9. Full name Paul [redacted]
10. Color or race Indian 11. Age at time of this birth ... yrs.
12. Birthplace Washington
13. (a) Usual occupation ...
(b) Industry or business ...
14. Residence ...
22. Mother's mailing address for registration notice:

MOTHER OF CHILD
15. Full maiden name Hahiasiat Kolockum
16. Color or race Indian 17. Age at time of this birth ... yrs.
18. Birthplace Washington
19. Usual occupation ...
20. Industry or business ...
21. Children born to this mother:
(a) How many other children of this mother are now living? ...
(b) How many other children were born alive but are now dead? ...
(c) How many children were born dead? ...

23. I hereby certify that the information was obtained from the Colville Agency Records and from personal inquiry. F. A. Gross, Supt. & S. D. A.
Chief Clerk's own signature
M. D., midwife, or other ... Date signed ...
Address Nespelem, Washington
24. Date received by local registrar OCT 30 1942
25. Registrar's own signature
26. Date on which given name added ... by ...
Form No. VS-81

Patrick's "Delayed Birth Registration" issued October 30, 1912. The year of birth is incorrect and should be 1892.

CHAPTER NINE

The Type Specimen

"The Patterson-Gimlin film did not bring a speedy resolution to the mystery of the sasquatch, as Patterson and others had optimistically anticipated. In fact, it made very little lasting impact on the scientific experts of the day, in the absence of a body or some bones."

—Jeff Meldrum, 2006.
Sasquatch: Legend Meets Science, page 21

Let's take a break from Patrick and look at the ecological and cultural context of Skanicum on the Colville and also review the typical morphology (form) of our apparently close relatives. This will fill-out Patrick's origin story and what his nuclear DNA (father's side) might have held.

Ed Fusch concludes his paper with an astounding story right from the Colville:

> Vince, a 30-year-old member of the Lake Band and a friend of this author, reports that as a small boy his grandfather, who was a Blue Jay, took him far up on [redacted] [name deleted for tribal privacy] Mountain to see Skanicum. He [Vince] was told

> to be very quiet and he would see them. He did see Skanicum close up and reports "he was Huge." He knows of a cave going deep into the mountain that is reported to be a home of Skanicum. He has been inside the opening which he states is squared and smooth, appears to have been worked on with tools, and was covered with tree branches. Inside is a strong Skanicum stench. He was afraid to go further into the cave.[88]

This account is remarkable to me on four different points. One is the reference to the grandfather being a "Blue Jay." Blue Jay birds among Native Americans in the Pacific Northwest are considered potentially sacred beings, capable of carrying prayers from Indians, or spiritual messages back to them. The unexplained reference to the Grandfather being a "Blue Jay" indicates that the grandfather was part of a group or society focused on spiritual pursuits. And somehow his Grandfather's status as one of this group gave him greater authority and facility in understanding Skanicum. This implies that, among the Lake Band anyhow, there is an assumed spiritual aspect to Skanicum.

The second insight that jumps out at me from this brief story: Some residents on the Colville in the last century knew and respected where Skanicum had "homes." Amongst certain circles at least, the presence of Skanicum was fact and routine to the point where a grandfather would take his small grandson specifically to see the "Huge" creatures.

Three: Specific caves are understood to be where Skanicum lives. Despite a general assumption in the Sasquatch enthusiast world that these hominids must use caves for shelter, almost no one reports signs of Sasquatch in caves, rock shelters, or in abandoned mines. It is odd that spelunkers never return with horrifying stories—or just disappear entirely. Instead, subterranean reports are very rare. But this story of Vince's is a firsthand account of someone returning from a cave convinced that it was actively used by Sasquatch.

Four: Vince reported that the entrance to the cave "appears to have been worked on with tools." That observation leads to the

[88] Fusch 1992, page 24

extraordinary proposition that Skanicum used metal tools to make a more satisfactory entrance to the cave. However, we know that white miners illegally crawled all over that reservation land looking for pretty rocks. More likely, miners found a natural cave and widened the opening a bit. Skanicum just opportunistically reinhabited the cave once the miners were gone or chased out.

Concealing the cave entrance with branches is very much in line with what we know of typical Sasquatch behavior.

The Colville is in relatively arid Eastern Washington. It would be easy to assume that this area would not be in Sasquatch territory. However, the reservation is largely in the southern reach of the rugged Okanogan Highlands, with rain-catching elevations of up to 6,774' (2,065 m), encouraging forests of deciduous and coniferous trees, year-round creeks and rivers, berries and tubers, and populations of elk and deer. Robert Michael Pyle writes in the 2017 edition of, *Where Bigfoot Walks: Crossing the Dark Divide*, "some of the most convincing reports of Bigfoot come not from dense forest but from the lavalands and shrub steppe east of the Cascade Crest."[89]

When I visited my mentor, Leatrice "Chick" Big Crow, on the Pine Ridge Indian Reservation of southwestern South Dakota in the early 2000s, she reported to me that Sasquatch sightings were becoming numerous there. Pine Ridge is drier, hotter, and has less forest cover than the Colville. It is even harder to explain Sasquatch on Pine Ridge than it is to justify them living on the Colville.

But perhaps there is something special about reservations that encourages Sasquatch activity? Pyle also says, "Great numbers of Sasquatch seemingly instinctively know that they are safer, perhaps accepted, in the confines of Indian Reserves and those surrounding lands."[90] Lapseritis takes it even further and suggests that it is a two-way exchange between Indian and Sasquatch, with the Sasquatch presence on reservations being beneficial and to "give counsel, spiritual messages, or warnings."[91]

It's my belief that Skanicum deserves to be treated as a

[89] Pyle 1995, 2017 edition, page 79
[90] Short 2024, page 34
[91] Lapseritis 2011, page 166-167

hominin. A hominin is by definition either a species of the genus *Homo* (which includes several recognized species such as *H. sapiens*) or of *Australopithecus* (bipeds found in Africa as early as 3 million years ago). I will sometimes use the looser term, hominid: Any member of the zoological family Hominidae, which includes all of the extant and extinct species descended from the common ancestor of humans (including *H. sapiens*) and the great apes, but *excluding* gibbons (the lesser apes). Living examples of Hominidae great apes include the panins (of the genus *Pan*: chimpanzees, Bili apes, and bonobos), gorillas, and orangutans. The use of *hominid* to describe unknown erect bipeds is a nod to the frequent perplexing reports of creatures that closely resemble bipedal examples of all three of these groupings of the great apes (such as the southeastern U.S. Skunk Ape reported sometimes looking much like an orangutan). You may run into the word *hominoid*. This is a very loose term that essentially says simply that the creature walked upright and had two arms, two legs, and a head with a face on it. Gray aliens are sometimes identified as hominoid. The term only claims anatomical resemblance, not relationship.

Note that I carefully used the phrase above, "deserves to be treated as a *hominin*." I'm not certain that Skanicum is a hominin, but functionally, they meet the requirements if for no other reason than this Patrick story. In Volume II, I will spend more time on the taxonomic identity of Sasquatch. This book has more than enough to explore in Patrick alone.

You may have heard or read about the need for a "type specimen" in establishing a species. A type specimen is essentially a parts of a preserved corpse, or even just a fossil, that is identifiable as representative of a distinct species, and may be studied to understand the unique physical characteristics of that species. Securing a type specimen generally allows for the naming of that species per recognized taxa. A taxon is a specific grouping of biological organisms by relatedness or degree of kinship. The recognized taxa (in the order of *increasing* genetic relatedness of its members) are domain, phylum, class, order, family, genus, species, and sub species. The taxa of the closely related humans that I am accustomed to dealing with (in decreasing order of relatedness) are sub species *sapiens*, species *sapiens*, genus *Homo*, Family Homindae (the great apes), Order Primates (yes, humans are primates), Class Mammals, Phylum Chordata (a reference to the spinal cord or vertebrae), and Domain Eukaryota (life with cells containing a nucleus).

Supposedly, the scientific world is refusing to recognize Sasquatch as a species wholly due to the lack of a body or body parts. I hesitate before criticizing scientists for not taking a firm stance on Sasquatch when they really have nothing solid to investigate. Biology deals with *things*, not ideas, and Sasquatch does not leave much in the way of *things* behind. I write this as someone who has for decades very much believed that there is something solid behind Sasquatch. But I still cannot say what Sasquatch is because *I do not know*. The Patrick story is potentially one of those great leaps forward in our understanding of Sasquatch because the hybrid *Patrick* was very much a *thing*.

On the other hand, in her thoughtful 2007 paper, *Scientific Inquiry into Bigfoot: Or How I Learned To Stop Worrying and Love the Sasquatch*, Anna Titcomb wrote, "In labeling the sasquatch as myth or fantasy purely based on the lack of a physical body, academics do themselves and the layperson a disservice in their blanket refusal to take into account secondary lines of evidence which may prove to be just as important."[92]

Before Patty looks closely at significant secondary lines of evidence.

Frustratingly, we occasionally hear rumors of what may have been a type specimen, usually whisked away by time, the elements, or mysterious government trucks and personnel. Perhaps one of the more annoying lost opportunities was that of the Minnesota Ice Man. In the fall of 1968, just one year after the Patterson-Gimlin film was taken at Bluff Creek in northern California, the eclectic biologist and author, Ivan T. Sanderson, was alerted to a carnival exhibit of an apparent hominoid corpse frozen in a block of ice. As it so happened at the time, Sanderson was playing host to the Belgian scientist, Bernard Heuvelmans, and together they immediately decamped to a Minnesota farm

---

[92] Titcomb 2007

where this frozen corpse was being held in the off season before its planned return to the carnival sideshow circuit.

Sanderson and Heuvelmans had two days starting December 17, 1968 to study, sketch, and photograph this ice man. [see Heuvelmans' sketch] At about six feet tall, Heuvelmans later described the male as "excessively and extraordinarily hairy" with dark brown hair three to four inches long "reminiscent of that of a chimpanzee." Yet the chest hair was relatively thin. The skin was the color of a white-skinned human corpse. The animal appeared to have been shot through the left forearm, with possibly the same bullet then entering the right eye, killing the creature. In Heuvelmans' sketch, we see a hominoid form with torso and arms disproportionately long compared to *H. sapiens* and legs relatively short. The eyes are high on the head with the implication of either a rapidly sloping forehead or a relatively small braincase (or both). The vertical distance between the bottom of the nose and the upper lip is also uncommonly large, along with the overall hand size, length of the thumbs, and the width, if not the length, of the feet.

B

From Heuvelmans 1969

Sanderson and Heuvelmans were convinced that they were examining an *actual* corpse. They could even detect the distinct smell of rotting flesh where the ice was thin.

Explanations of the origin of the corpse shifted in those first couple of years. The original story was that the body was found by Russian seal hunters frozen in a block of ice and floating in the Sea of Okhotsk. The next version had it found by Japanese whalers. In time, the Ice Man had simply been purchased in Hong Kong.

The possessor in 1968, Frank D. Hansen, said that he was "renting" the corpse since he had started showing it in May 1967. When the Smithsonian contacted Hansen in 1969 with the object of purchasing the specimen, Hansen said that the "owner" had taken the corpse away and that he would only be displaying a fabricated model in the future.[93] Indeed, firsthand reports from witnesses of the Ice Man shifted from spectators convinced that they were viewing an actual body, to disappointment at seeing a clumsy imitation.

Heuvelmans was so convinced that he had examined an actual type specimen that in 1969 he published a paper in French on the examination: *Preliminary Notes on a Specimen Preserved in Ice, of a Form Still Unknown of Living Hominids: Homo Pongoides*.[94] Heuvelmans had ventured assigning a taxon to the Minnesota Ice Man within the Homo genus: *Homo pongoides*. This name suggests a human that resembles the great apes (the pongids of gorillas, chimpanzees, and orangutans).

Relevant to our Patrick story, Heuvelmans writes [in my own Google Translate version]:

> One may still wonder whether it is legitimate to apply the classical criteria of specificity to Hominids. Let us not forget that these criteria are only accepted where crossbreeding occurs naturally, that is to say in nature, and not artificially, that is to say as a result of human intervention, as in hybridizations carried out in captivity on domestic or wild animals.

Heuvelmans is acknowledging that inclusion of the Minnesota Ice Man in *Homo* is only appropriate if *crossbreeding occurs naturally*. Was Patrick the product of such a "natural" crossbreeding?

Heuvelmans goes on to write, "of the six possible hypotheses, relative to the nature of the present specimen… the fifth (Man of a distinct subspecies) seems the most plausible and therefore the most worthy of being considered and retained." He further concludes, "The present specimen constitutes the holotype of the species bearing this scientific name." Heuvelmans was asserting

---

[93] Clark 1993, pages 215-217

[94] Heuvelmans 1969

that the Minnesota Ice man was the type specimen of *Homo pongoides*.

I discovered recently that Michael Thompson, another member of the Boreal Bigfoot Research Group—and co-founder of the twice held, furthest north, Boreal Bigfoot Expo in Fairbanks, Alaska—actually saw the Minnesota Ice Man *personally*. He writes:

> It would have been around 1974 or 1975. Maybe as late as 1976. I would have been 10 in the fall of 1974 and in Boise, Idaho. So it could have been the Idaho State Fair. I remember the exhibit had a sort of rabid-looking sasquatch painted on the outside and a guy wearing a cowboy hat and a woman with skimpy clothes. I remember going up three or four steps into a room and it was in the center.
>
> It smelled too... Think of walking into a walk-in freezer where a moose is hanging... It was in a row of other side show exhibits too... Two-headed pigs preserved in jars, etc. But the Ice Man was in this trailer by itself...[95]

Did we have a type specimen in the Ice Man and did it most properly belong in the human genus, *Homo*? Unfortunately, the Ice Man that Sanderson and Heuvelmans were able to examine soon disappeared—perhaps because whomever was responsible for its death and exhibition may have wondered if there might be legal repercussions, particularly if it was to be scientifically identified as a member of the human family. The simplest explanation is that whatever was frozen and rotting in that ice was a Sasquatch and that it had been shot somewhere in the north central United States. It's size would indicate a juvenile male, perhaps only six to ten years old.

In losing that singular holotype in the Ice Man, for whatever reason, we have gained fifty-five more years of mystery and argument. How would the Sasquatch world look now if the Minnesota Ice Man had become a type specimen in the *Homo*

---

[95] 11/17/2025 Michael Thompson, Sasquatch Tracker, from personal correspondence with the author

family in 1970? We would be openly sharing the continent with a wild people living in the wild places. We would not be alone and we would have an example in the Night People of true freedom and true self-responsibility. Everything about our world now would be different.[96]

Minnesota Ice Man placard ca 1970s, courtesy of Michael Thompson, Sasquatch Tracker

---

96 For more on the Minnesota Ice man see also Burtsev 2015, page 95; Debenat 2009, pages 242-248, 376; Redfern 2016, pages 181-183

Before Patty

CHAPTER TEN

What Does Skanicum Look Like?

Morphology is the form of things: The exterior package or how something looks to an observer. In the case of a living being, such as Skanicum, morphology is described in terms such as color, height, hair distribution, weight, skeletal shape, etc. I have heard that it is helpful to have a good idea of what Sasquatch looks like before encountering one in order to dampen the extreme terror most people initially report. Morphology might be worth our attention.

Albert Ostman, in his narration of what he encountered in 1924, gave a very detailed description of the hairy family that held him in central British Columbia:

> The girl would not hurt anybody. Her chest was flat like a boy—no development like young ladies…
>
> The young fellow might have been between 11-18 years old about seven feet tall and might weigh about 300 lbs. His chest would be about 50-55 inches, his waist about 36-38 inches. He had wide jaws, narrow forehead that slanted upward round at the back about four or five inches higher than the forehead.
>
> The hair on their heads was about six inches long. The hair on the rest of their body was short and thick in places. The

women's hair was a bit longer on their heads and the hair on their forehead had an upward turn like some women have—they call it bangs, among women's hair-do's.

Nowadays the old lady could have been anything between 40-70 years old. She was over seven feet tall. She would be about 500-600 pounds.[97] She had very wide hips, and a gooselike walk. She was not built for beauty or speed. Some of those lovable brassieres and uplifts would have been a great improvement on her looks and her figure.

The man's eyeteeth were longer than the rest of the teeth, but not long enough to be called tusks. The old man must have been near eight feet tall. Big barrel chest and big hump on his back—powerful shoulders, his biceps on upper arm were enormous and tapered down to his elbows. His forearms were longer than common people have, but well proportioned. His hands were wide, the palm was long and broad and hollow like a scoop. His fingers were short in proportion to the rest of the hand. His fingernails were like chisels.

The only place they had no hair was inside their hands and the soles of their feet and upper part of the nose and eyelids. I never did see their ears, they were covered with hair hanging over them. If the old man were to wear a collar it would have to be at least 30 inches.

...I was watching the young fellow's foot one day when he was sitting down. The soles of his feet seemed to be padded like a dog's foot, and the big toes was longer than the rest and very strong. In mountain climbing all he needed was footing for his big toe.

They were very agile. To sit down they turned their knees out and came straight down. To rise they came straight up

[97] **Author's note**: Christopher Murphy estimates in his paper, *Considering the Math—Patty's Height,* that Patty was 7'-4" tall and 700 pounds! [97] And Patty was a female of a species in which the males generally are *significantly* larger than females, just as with *H. sapiens* and with all of the *Homindae*. Murphy, https://www.sasquatchcanada.com/uploads/9/4/5/1/945132/considering_the_math_--_revised.pdf, and Meldrum 2006, page 177

without help of their hands and arms.[98]

Sasquatch often appear very human in form, part of the puzzle of their origin. In Steve Isdahl's 2020 book, *The Day Sasquatch Became Real For Me*, he quotes a Michigan man as writing that a young one, "looked like a young black child with no clothes on."[99] In Paulides' summary he said that he had found that the northern California Sasquatch around, and in, Indian reservations, often had human-like facial features.[100] And Scott Marlowe in his 2013 book, *Bigfoot Enigma*, observes, "If you were to strip the hair from the North American Bigfoot, it would likely appear more human than ape."[101]

Locomotion:

Locomotion in Sasquatch is usually bipedal, similar to humans, but occasionally they are seen moving quadrupedally—either knucklewalking like a gorilla, or sometimes they use their hands to grasp the ground and pull themselves forward with great effect and speed. Only a few witnesses have seen Sasquatch transition from two legs to four or vice versa. As Dr. Henner Fahrenbach wrote in his 1997-1998 paper, *Sasquatch: size, scaling, and statistics*, "Bipedal gait, seemingly as efficient as a quadrupedal gait (Rose 1984), can be viewed as an adaption to becoming an endurance hunter in the very demanding terrain inhabited by the Sasquatch."[102] Meldrum elaborates:

> "...the evolution of bipedalism is not a singularly human trait, but a manner of moving on the ground that has evolved independently in parallel, multiple times. Indeed, apes display a clear tendency to adopt an upright posture and to walk on two feet from time to time—referred to as facultative bipedalism.

---

98 Bigfoot Field Researchers Organization Report #1091, https://www.bfro.net/GDB/show_report.asp?id=1091
99 Isdahl 2020, page 181
100 Paulides 2017, pages 451-452
101 Marlowe 2013, page 75
102 Fahrenbach 1977-1978

> It appears that more than one species of ape became a full-time biped, or *habitual* biped.[103]

It appears that Sasquatch is a *facultative quadruped*—some go to all fours when situations call for it (for stealth, for keeping to cover, for misdirection), but they are usually most comfortable covering ground on their own two feet. I suspect that Ostman's description of how his hairy captors sat down is a clue into how their pelvises are structured to accommodate four-legged movement: "To sit down they turned their knees out and came straight down."

Fahrenbach felt that walking upright was the Skanicum's preferred mode of getting around, "Despite the fact that efficiency of locomotion rises substantially with speed (Heglund 1985), such running as occurs in the sasquatch seems to be restricted to short bursts of speed to reach cover, during which the metabolism for a sasquatch could rise approximately ten- to twenty-fold (Jungers 1985a), and hence, would be used sparingly."[104]

The strongest examples of tracks that I have seen (partly melted in half an inch of snow on the thick ice of a frozen back country stream in Alaska in January) confused me because they were in a single line (not staggered from side-to-side) and about four feet from the toes of one foot to the toes of the next foot. I kept looking for the imprints of the foot on the other side of its body before I realized that the long stride and the in-line aspect were all that I was going to find. Paulides and others refer to this type of trackway as *tightrope walking*.[105]

Baxter writes:

> It's theorized that, while Bigfoot walks bipedally like us, they have different shaped hips which causes their footfalls to appear to be in a straight line, not staggered, like human tracks.[106]

---

[103] Meldrum 2006, page 223
[104] Fahrenbach 1977-1978
[105] Paulides 2017, pages 159-169
[106] Baxter 2021, pages 88-89

The slouching walk of Patty looks odd to us. I can recall only one person that typically walked with knees partially bent rather than the way we normally touch down—with legs fully extended, as if we are on stilts. Now that an old knee injury causes me problems, I find that I sometimes imitate this bent-knee walk, particularly when I am carrying loads. This accommodation reduces the shock to my knees and hips with the cushioning of bent joints. This approach to walking is called a *compliant* gait. Meldrum confirmed its efficacy in helping me deal with a bum knee, "...a compliant limb posture alone may reduce stresses by as much as 18 percent."[107]

Fahrenbach defines a *step* in relation to studies of Sasquatch trackways as, "…a point on one foot to the identical point on the opposite foot…"[108] Such as from the rear of the heel of one foot to the rear of the heel of the next. Fahrenbach goes on to expand upon the description of a compliant gait:

> The sasquatch has evidently evolved a peculiarly long stride with a long arm swing, a slow cadence coupled with prolonged ground contact of the feet (Krantz 1992), and a suppression of vertical body oscillation, a so-called compliant walk (Alexander 1977). This bent-hip, bent-knee bipedal gait is deviant from the human stiff-legged mode, and has frequently been commented upon by eyewitnesses for its conspicuous fluid grace…[109]

Grossinger points out that tracks indicate that Nahganne almost always walk flat-footed.[110]

Pulling it all together, Meldrum summarizes why the Sasquatch locomotion we see in the "trace evidence" of tracks, and in the direct evidence of the Patty video, makes perfect sense:

> The sasquatch appear to have adapted to bipedal

[107] Meldrum 2013 page 218
[108] Fahrenbach 1977-1978
[109] Ibid.
[110] Grossinger 2011, page 59

> locomotion by employing a compliant gait on flat flexible feet. A significant degree of prehension has been preserved in the toes by retaining the uncoupling of the propulsive function of the hind foot from the forefoot by means of the midtarsal break. The toes are spared the peak bending forces experienced by human toes at toe-off by means of the compliant gait with its extended period of double support (the interval of a gait during which both limbs are in contact with the ground and support body weight) and push-off from the front half of the foot. All in all, this would be an efficient strategy for a giant terrestrial bipedal ape. In fact, it is an extension, more likely in parallel, to the very strategy that marked the mode of bipedalism used by early hominids. …eg., increased breadth and further lengthening of heel, to accommodate its gigantic body proportions.[111]

Meldrum almost slips an important point past us in the above passage when he writes, "…an extension, more likely in parallel, to the very strategy that marked the mode of bipedalism used by early hominids." When he talks about early hominids, he's referring to a time at least three million years before present (bp). What he suggests here is that Sasquatch is unlikely to actually be, or to be evolved directly from, these early hominids, but that Sasquatch may have evolved later and *parallel* to these early hominids—*independent* of them. *Gigantopithecus* remains have been dated from times much more recent than those of the early hominids, perhaps within the last 100,000 years. Meldrum may have revealed in that last passage his serious consideration of a Sasquatch origin from *Gigantopithecus* or possibly from even more complicated beginnings.

Torsos:

Meldrum explains that the well-developed muscles on the shoulders and back of the neck attach higher on the base of the skull in Sasquatch than they do in modern humans, giving an adult

[111] Meldrum 2006, page 247

Sasquatch the classic no-neck stooped appearance.

The other reported characteristics of their torsos reveal some interesting things. Meldrum again, studying the anatomy visible in the Patty film, debates if the lack of a visible gut—a protruding gut such as seen in gorillas—implies that their intestinal tracts are not adapted for the fermentation of coarse vegetation. Alternatively, the relatively long torso may be able to contain that large gut: "...the generalized anatomy seems consistent with a moderate-length gut combined with a generalized omnivorous diet."[112] And, "Descriptions of sasquatch seem to lack mention of a noticeably protruding belly characteristic of a dedicated herbivore.... The broad shoulders, deep torso, and short flaring pelvis of a giant biped would accommodate a large abdomen less noticeably."[113] From a morphological perspective, it would appear that a diet heavy in coarse vegetation *may* be possible for Skanicum.

Despite what Ostman had to say about the flat chest of the young female that he encountered, witnesses often report very noticeable breasts on females, sometimes quite large.[114]

One of Alley's witnesses reported that a seven to eight foot tall female Sasquatch at the shoreline of southeast Alaska appeared heavily pregnant, bulging from her hips to just under her breasts ("about the size of dinner plates").[115] Imagine Patty with child and weighing perhaps 800 pounds (360 kg)!

I cannot recall one witness observation of genitalia on a female Skanicum, but penises are sometimes observed when hair cover allows. Rarely, penises have been observed erect or semi-erect. With few exceptions, penis size is not reported as proportionally larger than that of the human male.

If you have a copy of Meldrum's thoughtfully researched 2006 book, refer to the comparison of human male and female figures next to Patty (that Patty image courtesy of Igor "$200" Burtsev) on page 169. This illustration shows Patty and the modern humans from the rear.

---

[112] Ibid., pages 167-168
[113] Ibid., page 188
[114] Paulides 2017, pages 406, 418
[115] Alley 2003 (2018 edition), pages 72-73

Speaking of the rear, did you know that there have been two Sasquatch buttock imprints casted? In May 1993, Paul Freeman discovered and cast a likely female buttock imprint along Dry Creek east of Walla Walla Washington, which measured 14-1/2" wide across the beam. By comparison, my light frame would have left an impression only about 10" wide. Meldrum observed:

> The cheeks of the buttocks were prominent and apparently well muscled. ...Its consistency with eyewitness observations of prominent buttocks on sasquatch is noteworthy. The gluteus maximus is a powerful retractor of the hip, important in climbing steep mountainsides, as well as holding the torso erect over the lower limbs. Its apparent development in a giant bipedal ape is appropriate to its posture and habitat.[116]

Meldrum goes on to describe the very similar anatomical finds in a September 2000 cast of what appears to have been a partially reclining Skanicum near Skookum Meadows in southern Washington that includes the left forearm, buttocks, thigh, and both heels:

> The cast reveals another hallmark of bipedalism–a pronounced buttocks, produced by a short, flaring hipbone and well developed gluteal muscles. The gluteus maximus in particular is a powerful retractor of the hip, a particularly important function when walking up steep inclines, characteristic of mountainous terrain. It also aids in keeping the torso upright over the supporting legs when in an erect posture. The presence of the pronounced buttocks in the skookum cast is correlated with the Achilles tendon, combined indicating an adaptation to bipedalism in this hominoid.[117]

Note in the last sentence the careful use of the word "hominoid," meaning a creature that resembles the human form and is a facultative or *obligate* biped (primary locomotion using

---

[116] Meldrum 2006, page 112

[117] Ibid., page 122

the two rear legs only). The choice of the word "hominoid" leaves open the possibility that the impression-maker is not necessarily human (of a *Homo* species) or even of the family Hominidae (which includes humans, chimpanzees, gorillas, and sometimes orangutans). Meldrum's careful parsing of words is befitting of an anthropologist without benefit of a type specimen to examine, but this book makes the equally careful case that we *are* dealing with a species suitable for inclusion in the *Homo* family—a qualifying *hominin*.

In case it hasn't been clear in the descriptions so far, even an adult female Sasquatch is far more robust than the most robust of modern human specimens. On page 177 of Meldrum's 2006 book, he offers a table to compare Patty's calculated dimensions to that of Arnold Schwarzenegger at his body-building peak. I took the comparison a step further and normalized Patty's measurements to that of the body-builder's height of 6'-2", shrinking her to that height and her measurements down by the same factor. The results were impressive. Although the chest circumference of the scaled down Patty was *exactly the same* as Schwarzenegger's, the bicep circumference was 123% of Schwarzenegger's, the calves 125%, the thighs 137%, and the waist 168% (the same numerical circumference as her chest at 57" or 145 cm!).

Her diminished weight was slightly more difficult to calculate. I used a formula by Dr. Grover S. Krantz published in *Northwest Anthropological Research Notes* in 1972.[118] Krantz points out that for similarly proportioned bodies, the volume (or weight) changes by the cube of the ratio of the change in length (or height). Shrinking Patty down from 88" or 7'-4" tall (Meldrum's figure) to Schwarzenegger's 74" is a change in height ratio of 74"/88" or 0.84. Cube that ratio change to get 0.5946. Multiply Patty's originally calculated weight of 700 pounds by 0.5946 to get 416 pounds if shrunk to 6'-2"! An impressive 177% of Armold's barbarian weight of 235 pounds. Keep in mind that this is a *female* of the Skanicum species. Males might be expected to be bulkier and more muscular yet, however not all Sasquatch are as stocky as the "Patty-type."

[118] Krantz 1972b, page 230

Some Sasquatch are so heavily muscled that it may be that they have reduced or absent myostatin, a protein that normally inhibits muscular growth to conserve calories for other uses in the body. Some humans have a genetic defect that causes reduced or absent myostatin. Those people end up comparatively heavily-muscled, even as toddlers. Some domestic animals have been bred for low myostatin to increase their meat production. Perhaps some Sasquatch are naturally low in myostatin or that breeding or group dominance will trigger the condition and result in extraordinarily-muscled individuals.

Danny Vendramini pointed out in his 2009, *Them & Us: How Neanderthal Predation Created Modern Humans*, that "...the more vertically the lumbar vertebrae are aligned, the more efficient they are at transferring stress, resulting in less injuries."[119] And a reconstruction of Patty's skeleton completed by Rueben Steindorf shows just this stacking of vertebra.[120] If humans had this same stacking we would expect to have fewer back injuries. However, one reconstruction of a Neanderthal spine (Been et al 2017), finding it much straighter than those of *H. sapiens*, reported that:

> Neanderthals might have been better adapted to carry heavy loads and, potentially, to engage in generally more rigorous upper body activities. On the other hand, it suggests that Neanderthals potentially had shorter stride length and slower walking velocity on flat terrain in comparison with modern humans.[121]

This Neanderthal spinal stacking came with the bottom of the sacrum rotated forward (the sacrum is the fused vertebra at the base of the spine) which would have reduced both the protrusion of the buttocks and the abdomen relative to humans. This rotation of the spine and pelvis might help to explain how Sasquatch can carry a large gut less noticeably than a gorilla. In contradiction to

[119] Vendramini 2009, page 135
[120] Meldrum 2006, page 173
[121] Been et al 2017, pages 248-249

Been et al's Neanderthal findings, Sasquatch seems to have a more vertically-stacked spine along *with* longer strides and a faster walk. There must be other compensatory anatomical adjustments in Sasquatch to account for their agile locomotion such as the more effective foot lever that Krantz observed[122] and their compliant gait.

One last point from Fahrenbach to make about the extreme size of the Sasquatch torso: "Most adaptive for the Sasquatch is the increased resistance to cold, since the radiative body surface increases with the square of the linear dimensions, whereas the heat-generating mass increases at roughly the third power."[123] The result being: The bigger they are, the less affected by cold.

To summarize some likely suppositions we can make based on Sasquatch torso morphology:

1) The torso is longer proportionally than those of humans—sufficiently longer that it may conceal the volume needed to ferment roughage.
2) What we see in the Patty video, and in the two buttock imprints cast, is an animal heavily-muscled for bipedal locomotion and an erect posture.
3) This is an animal with a short pelvis (more similar to the pelvic girdles of *Homo* than the relatively tall structures of gorillas).
4) Sasquatch likely have vertebrae more nearly vertically-stacked than the gentle s-curve of *H. sapiens*.
5) With all due respect, the build of the female Sasquatch torso, in Patty's case at least, is similar to that of a 55-gallon drum in dimensions and shape.
6) The large mass of the Sasquatch helps to maintain body warmth in cold climates and allows it to range further northward sans clothing than other *Homos*.

Arms & Legs:

[122] Krantz 1972b, pages 233 & 235
[123] Fahrenbach 1997-1998

Among *H. sapiens*, some Australian Aboriginal populations have been found to have the longest legs relative to torso length, while some rural Guatemalan peoples have the shortest legs.[124] Similarly, there seems to be significant variation in Sasquatch limb to torso length ratios. Patty has somewhat proportionally shorter legs, but her arm length relative to her torso does not appear remarkable. But some witnesses, such as of Skunk Apes, report fingertips hanging below the knees. On the other extreme, Alley has a 1955 Southeast Alaska report of a ten-foot-tall (3 m) Sasquatch with "long skinny legs." The associated 19" (48 cm) tracks support the height claim.[125] See Table 1, page 110.

Another second hand Alley report is of a skeleton being found on the ground within an enclosure of logs, two high, stacked "corral-style." The skeleton was reported as "eight or more feet long." The length of the leg bones were particularly memorable.[126]

Another thing to keep in mind when it comes to limb length is what Bogin and Rios pointed out in their 2003 paper, *Rapid morphological change in living humans: implications for modern human origins*, while referencing a much earlier work:

> At birth, head length is approximately one-quarter of total body length, while at 25 years of age the head is only approximately one-eighth of the total length. There are also proportional changes in the length of the limbs, which become longer relative to total body length during the years of growth (Scammon 1930).[127]

Practically, this would mean that younger Skanicum/Sasquatch would appear proportionally to have relatively normally human-proportioned limbs and larger-proportioned heads than they will at adulthood. That might explain why my possible sighting of a 5'-6" (1.68 m) black figure in the gloomy woods of Illinois did not strike me as having anything but normal human proportions.

---

[124] Bogin and Rios 2003, page 74

[125] Alley 2003 (2018 edition), pages 90-91

[126] Ibid., page 264

[127] Bogin and Rios 2003, page 72

According to Fahrenbach, and my chart on page 110 (derived from various sources), this individual would have been about ten years old and weighed somewhere in a range of 130 to 200 pounds (59-91 kg), depending on body-type.[128]

Burtsev reports that in one area, Almas children would reach "maturity" at about ten years of age and then leave the clan, some returning later.[129] Other indications of growth rates and age include the claim that a one-year old Sasquatch will be the size of a two or three-year old *H. sapiens* child[130]; a two year old was reported at about four feet tall[131]; and another in British Columbia, Canada, known to be only seven years old, was nearly seven-feet tall.[132]

Heads, Skulls, & Faces:

Sasquatch heads, particularly in adult males, are often reported as pointed or cone-shaped.[133] This cone-shaped effect may be a sagittal ridge or sometimes a more pronounced sagittal *crest* that forms to provide the attachment area for strong chewing muscles, such as is typically found in adult male gorillas. The sagittal crest "...is really a function of the ratio of body mass to cranial volume and the robustness of the jaws."[134] Meldrum continues to point out that larger female gorillas may also develop a reduced sagittal crest. I found out when moving sheets of plywood many hundreds of feet by trail, balanced on the top of my own head, that I needed skull padding because I have a poorly developed sagittal *ridge* or keel that was taking an uncomfortably concentrated load.

The 5'-6" tall, apparent young Sasquatch whom I sighted in Illinois, did not have a noticeable sagittal crest or cone head. Patty,

---

128 Fahrenbach 1997-1998, page 8
129 Burtsev 2015 page 173-174
130 Lapseritis 1998, page 49
131 Noël 2019, page 38
132 Burtsev 2015 page 369
133 Bigfoot Field Researchers Organization Report #1091 (Ostman account), https://www.bfro.net/GDB/show_report.asp?id=1091; Isdahl 2021, pages 28, 29, 153, 211, 247; Noël 2019, page 100; Paulides 2017, pages 290
134 Meldrum 2006, page 166

however, does have a pronounced peak rising to the back of her skull. This ridge may be a feature that develops in individuals only after years of chewing and may be entirely absent on some adult males if dietary conditions do not call for the development of securely-anchored large chewing-muscles. This lack of need might be the case with more omnivorous or carnivorous diets.

To explain the typically stooped posture of *adult* Skanicum, a reasonable speculation is that the *foramen magnum* (the opening in the base of the skull that the spinal cord passes through)[135] may be shifted slightly towards the back of the skull relative to those openings in "modern" humans. This positioning would cantilever the center of mass of the head forward, rather than centering the mass over the *foramen* magnum and the spine as in modern humans. This positioning would require more developed neck muscles to support the head, as is frequently described by witnesses. With the cranium shifting down and rearwards on the Skanicum, the face would naturally tilt upwards (without significant structural accommodations). As compensation, a hunched-over positioning of the shoulders keeps the face pointed forwards, rather than up.

This shifting of the *foramen magnum* backwards on the skull base may be something that happens largely *after* birth. It is possible that this feature is in about the same position in Sasquatch and *H. sapiens* children at birth, but the relative position changes as the brain develops. From *Foramen Magnum Placement*, an online article by Melanie Beasley, Beasley points out that, in some primates, this skull opening indeed shifts towards the rear of the skull as the young develop:

> As human infants grow, their foramen magnum remains in the center of the skull, while posterior drift of the foramen magnum occurs in nonhuman primates through the resorption at the posterior end of the cranial base and deposition at the anterior end. Accordingly the position and orientation of the foramen magnum is considered to be a reliable reflection of mode of locomotion in fossil hominins. The foramen magnum

[135] Roberts 2018 edition, page 69

position has been cited as evidence of bipedal locomotion (a hominin characteristics [sic])...[136]

In quadrupeds, this opening is at the rear of the skull to accommodate forward-facing when the spine is parallel to the ground. In *H. sapiens,* the *foramen magnum* is centered under the base of the skull to facilitate our erect posture. In Sasquatch, this opening may be shifted slightly towards the rear, but not to the same extent as an *obligate* quadruped, but resulting in the typical stooped posture of Sasquatch. This *foramen magnum* positioning may be transitional between quadrupedal and bipedal movement. For the Skanicum, this positioning would make it easier to look forwards when on all fours, the back and shoulders now very straight or the back perhaps even concave with the chest lifting through the shoulders to bring the face up.

The delayed development of certain physiological features, such as the *foramen magnum* shifting towards the rear of the skull base as the brain grows and develops, may help to explain why young Sasquatch, like my curious fellow in the woods of Illinois, seem so human-like in form, while the adults may appear so different.

Meldrum describes the braincase volume of Sasquatch as smaller proportionally than in humans.[137] That was certainly the case in the video of the possible human/Yeren hybrid and initially *appears* to be the case in Patrick. This smaller proportion does not necessarily mean that the brain of an adult Sasquatch is actually *smaller* than a typical modern human brain. An adult male Skanicum might have five times the total body volume of an adult male human, but only have a brain two or three times as big. *That's still a very large brain*. Monson et al point out that there is a strong proportional link between body size and cranial volume in Hominidae. Consistently, the bigger the body, the bigger the brain.[138] We will revisit this brain volume discussion and its effects when we look closely at Patrick's cranial profile in Chapter

[136] Beasley [accessed 2026.02.01]
https://carta.anthropogeny.org/moca/topics/foramen-magnum-placement

[137] Meldrum 2006, page 167

[138] Monson et al 2024, page 5

16.

Facial form and the underlying structure seem to vary widely. Many witnesses naturally fall back to describing their own concept of a caveman or Australian Aboriginal.[139]

In the case of Patty, Meldrum describes her jaws as projecting slightly beyond her flattened nose, but that she was not nearly as prognathic (in which the lower jaw extends forward beyond the rest of the face) as chimps or gorillas. He suggested that this flat face is consistent with a coarse diet and would shift bite loads to enlarged premolars and molars. He found her lower face relatively long, with the lower jaw flaring towards the back of her skull, providing greater attachment area for the enlarged chewing muscles in front of her ears.[140]

Significantly, Sasquatch is reported sometimes with very prominent and strong chins.[141] As Vendramini points out, "...the human chin is unique among mammals."[142] Gorillas and chimps do not have chins (the *mentum*—a forward-projecting bone at the front of the lower jaw). Their jaws recede rapidly under the lower teeth. The presence of chins in Sasquatch is yet another reason for us to not quickly leap to describe them as apes. Unfortunately, Patrick's profile will throw a monkey wrench into my reasoning about chins.

Brow ridges are typically reported from minor to extremely pronounced, to the point of plunging the eyes into dark obscurity.[143] Alley suggested their night vision must be two to three times the acuity of our own.[144] This estimate may be on the low end.

The nose is generally described as wide and flat, but hooded, despite popular demeaning caricatures of Sasquatch with ape-like forward-facing nostrils.[145] Sometimes the nose is described as a

[139] Burtsev 2015 page 176; Short 2024 pages 132

[140] Meldrum 2006, page 167

[141] Grossinger 2022, pages 153, 190

[142] Vendramini 2009

[143] Burtsev 2015 page 176; Grossinger 2022, page 191; Isdahl 2020, page 143

[144] Alley 2003 (2018 edition), page 299

[145] Isdahl 2020, page 247; Marlowe 2013 pages 25-26; Meldrum page 167; Paulides 2017, page 370

boxer's nose.[146]

Ears, rarely seen through the hair,[147] are reported as human-looking, but somewhat smaller, proportionately, than human ears.[148]

Hair:

Hair patterns are also extremely variable. Patty is on the more hirsute side, male or female, of what is normally reported. Generally, there are areas free of hair at least on a low forehead and around the eyes and nose.[149] Often the chest, in contrast with human males, is just thinly haired. But there have been balding male Sasquatch[150] and others so thinly haired overall as to more closely resemble a hairy modern-human man. Noticeable brows and even uni-brows are sometimes reported.[151]

Hair color is most often described as black or a dark brown, with shades of auburn or even reddish-brown coming in next. White, gray, and dark hair with a bluish hue are all reported less often. Occasionally one will be spotted with irregular patches of contrasting hair or even stripes down the back or across the chest. Michael Thompson of Sasquatch Tracker finds this "barring" of light hair color to be a significant pattern in his Interior Alaska reports.

Hair patterns often include longer hair hanging off the outer edge of the forearm almost like a buckskin fringe. Sometimes hair thickens and lengthens into a beard, but those full beards may not start until below the jawline. Paulides suspects a pattern of hair darkening with age.[152] Grossinger has a report of long head hair continuing down the upper back along with a moustache on the

---

[146] Grossinger 2022, page 153; Isdahl 2020, page 211
[147] Grossinger 2022, page 190
[148] Isdahl 2020, page 153; Marlowe 2013, pages 25-26 (1955 William Roe encounter near Valemont, British Columbia
[149] Isdahl 2020, page 247; Paulides 2017, page 370
[150] Ibid.,, page 28
[151] Ibid., page 143
[152] Paulides 2017, pages 351

same creature.[153]

Body hair may be patchy and thin or consistent in length and covering from only about 1" (such as in the 1955 William Roe account in British Columbia) to up to 12" or more. Vendramini points out that "...thermal stress is up to three times greater on a naked human than on a primate covered with terminal hair. …the fur of savannah primates is actually denser than that of forest-dwelling primates."[154] So we would not necessarily expect to find longer hair on farther north Sasquatch. Long hair actually has an insulating effect against *heat*.

Suspected Sasquatch hair samples have puzzling properties in structure and in genetic traces. Mammal hair typically has two or three components. The *medulla* is the hollow core of hair–made of air spaces–which may be continuous; or the air cells may be discontinuous, but distributed in a pattern; irregularly discontinuous; or the medulla may be absent altogether (no hollow core at all). The medulla is surrounded by tough keratin (types of tangled, fibrous proteins also found in fingernails and toenails and the outer skin) which makes up the *cortex* (the inner core of hair minus the voids of the medulla, if present). It is this cortex which carries the bulk of the hair coloration. The cortex is wrapped by a final layer of overlapping keratinized scales called the *cuticle*. The color and the makeup of the medulla, cortex, and cuticle are what give each species their own distinctly identifiable hair.[155]

Unfortunately, labs have had a difficult time sequencing Sasquatch DNA from both hair shafts and follicles. Possible explanations given include: "the small diameter of the hair combined with the lack of a cellular medulla, the duration and conditions of the storage of the hair before analysis, or some species-specific property of the DNA."[156] The last suggestion of a "species-specific property of the DNA" caught my eye, but apparently gorilla hair is difficult to amplify as well.

Alley wrote of the twelve hair samples that he had collected,

---

[153] Grossinger 2022, page 88
[154] Vendramini 2009, page 130
[155] Bryant & Trevor-Deutsch 1980 pages 297
[156] Meldrum 2006, page 265

"all have a reddish tinge under the microscope, and all are lacking a medulla."[157]

Meldrum did collect enough reasonably likely Sasquatch hair samples that he was able to build a profile of expected hair characteristics:[158]

- Variations in color and thickness along the length of individual hairs (human hair does not share the extent of these variations)
- Light in weight
- A human-like scale pattern
- Absent or lacking a continuous medulla core (normally a medulla will only be found in large, thick Sasquatch hairs)
- Related to, but distinguishable from, the human-chimpanzee-gorilla group
- More like gorilla hair than human or chimp hair
- Pigmentation being extremely fine and diffuse
- Uncut ends, even in short hairs
- Cuticle (outer layer) pigmentation irregular in pattern
- Occasional split ends
- Average length less than 10"

<u>Teeth:</u>

Teeth are often described as big (sometimes oversized), square, and even.[159] The teeth have been compared to brilliant white chiclets (a uniformly square candy-coated chewing gum, surprisingly refreshing), but range considerably in appearance with one witness describing, "dirty off-white greyish teeth, which were very human-looking."[160]

Mills et al examined teeth marks found on elk bones and were able to measure the likely width of one tooth that had penetrated

---

[157] Alley 2003 (2018 edition), page 260
[158] Meldrum 2006, pages 263-264
[159] Isdahl 2020, page 247
[160] Grossinger 2022, page 153

into the elk bone. From the marks on the bone, they were able to deduce that the piercing tooth was 0.55” (13.97mm) wide and 0.15” (3.81mm) thick.[161] By comparison, my widest tooth is about 0.33" wide. Assuming tooth width proportional to height, this ratio of Sasquatch tooth width to my own (55/33) multiplied by my height (6’ or 1.83 m) would suggest that the elk bone chewer towered to 10’ or about three meters tall!

Canines reported by witnesses range from non-existent to significantly enlarged.

In tellings of his 1955 sighting, William Roe described the female Sasquatch as using its even teeth to pull leaves off branches for eating.[162] I cannot recall an account of missing or badly formed teeth on a Sasquatch. Which is odd.

Eyes & Face:

Eyes are generally reported as black or brown, but there are a couple reports of eyes all or part *orange*.[163] There are also consistent reports of blue eyes.[164] Blue eyes beg the question: Is this blue eye color a sign of Neanderthal admixture into Sasquatch as it is in humans? Or, more to the point of this book, is this a sign of human admixture into Sasquatch? The whites of the eyes (the *sclera*) are sometimes absent or greatly diminished from the human norm.[165]

Paulides (and others) report that his tribal witnesses in northern California describe facial appearances very close to those of humans, despite the very hairy Patty filmed not very far away: “peaceful and human in appearance.”[166]

Skin:

---

[161] Mills et al 2015, page 26
[162] Marlowe 2013 pages 25-26
[163] Paulides 2017, pages 421 and 422
[164] Burtsev 2015, page 175; Isdahl 2020, pages 59, 211, 225; Lapseritis 1998, page 46; Short 2024 page 246
[165] Isdahl 2020, page 211
[166] Isdahl 2020, pages 82, 83, 143; Paulides 2017, pages 201, 236, 268, 274

Skin color ranges from (rarely) typical Caucasian white, deep tan, gray[167], to black, with the greatest number of reports falling into a shade of charcoal.

There are rare reports of persistent greasy hand and face prints being left at the scene of Sasquatch activity such as what Christopher Noël describes in his 2019 book, *Mindspeak: Tapping into Sasquatch and Science.*[168] I will suggest a possible explanation for that greasy observation in a later chapter. Noël described finding oily/dirty fingerprints on auto glass. The prints were clear enough that he noticed two odd things about the fingers: The fingers tapered towards the ends more than a human's and the fingers had side fingerprint "striations" unlike in humans.[169]

Hands:

There are large variations in the proportions and structure of Sasquatch hands. Very human-like in Oregon;[170] however Meldrum's careful analysis of hand and finger casts led him to observe that the fingers are short in proportion to those of modern humans (likely due to the fleshy webbing between the fingers extending further towards the second finger knuckle from the fingertips than is typical in modern humans). And he found that the thumb is rotated more towards the orientation of the fingers (less readily opposable for grasping). Finally, he noted that the muscles at the base of the thumb in the palm are less developed than in humans.[171]

At least three of Grossinger's witnesses observed (and the Patty film confirms) that the palms tend to face rearward when walking, creating a loose flapping effect at the extremes of reach, both backwards and forwards.[172] The norm for a slim modern human is for the palms to face towards the thighs with less

---

[167] Isdahl 2020, page 211; Noël 2019, pages 100
[168] Noël 2019, page 264
[169] Ibid., page 17
[170] Isdahl 2020, page 124
[171] Meldrum 2006, pages 107-108, 110
[172] Grossinger 2022, pages 153, 189, 227

movement of the hands out of line with the forearms.

Witnesses are often struck by just how large the hands are. One person writing to Isdahl reported hands the size of dinner plates and the fingernails yellowish.[173] Paulides quotes a witness as describing the hands as "huge in comparison to a human's."[174] Not surprisingly, the hands may develop thick pads on the underside of the palms and fingers, much like what has been reported on the feet of Sasquatch.[175]

Tracks & Feet:

The tracks that I saw at about 13" long were unusual in that they had a slightly hour-glass shape. Generally, suspected Sasquatch tracks are flat-footed (no arch) and often almost as wide at the heel as they are across the ball of the foot, imparting a rectangular block appearance to imprints. In a statistical analysis of 706 tracks from different individuals, Fahrenbach found the average length of the foot track was 15.6" (16" for the *median* length), with adult foot sizes ranging from 13" to 19".[176] Keep in mind that shorter tracks are likely under-recognized and under-reported because of the possibility that they are those of barefoot Hairless Ones. This reporting feature may inflate average track size.

In 1980, George W. Gill, in his paper, *Population Clines of the North American Sasquatch as Evidenced by Track Lengths and Estimated Stature*, observed that there was a rise in reported Sasquatch track length with increasing latitude, as would be expected if Sasquatch was subject to Bergman's Rule (that the average body size within warm-blooded species increases in cooler high latitudes versus warmer areas in lower latitudes).[177] To my knowledge this thorough of a statistical analysis of track sizes has not been repeated since.

Back in 1972, Krantz pointed out that, "An 8 ft. tall, heavily

[173] Isdahl 2020, page 124
[174] Paulides 2017, pages 403-404
[175] Noël 2019, page 100

[177] Gill 1980, pages 267-268

built hominid would require certain structural modifications in its feet because of its great absolute body weight." He suggested that the weight-bearing surface of the talus (the ankle bone) would be proportional to weight load in any biped or quadrupedal animal and that something as large as a Sasquatch would have correspondingly broader heels to accommodate this large talus surface. The average breadth of Sasquatch heel that Krantz was aware of was 5-1/2". According to his calculations, this corresponded to an average weight of 760 pounds. However, he also noted that weights, "…in excess of 1,000 lbs. can probably be ruled out as being beyond the weight-bearing capacity of the ankle."[178] I suspect that there must be some further structural compensations to allow for greater weights than Krantz thought possible. Meldrum's "extended period of double support" of two-foot ground contact, or the weakly-evidenced theory of "super-strong" bones, or the shock-absorbing effect of a thick fatty-layer on the undersides of Sasquatch feet and toes are some possible compensations for ankle weight-capacity. By Krantz's calculations, an 11' Sasquatch of Patty's build would weigh over a ton. However, occasional eleven-footers (and beyond!) are reported by seemingly sincere and careful witnesses.

In the large number of tracks that he was able to study, Meldrum found that the maximum width to length ratio of these foot impressions was somewhat consistently 0.5.[179] He also noted "...a disproportionate length of the toes and heel by comparison to the human foot."[180] Meldrum found that the reported proportions and size of a Sasquatch track were to be expected: "Given its extreme size and weight, the sasquatch has an even more pronounced heel and a broader and well-rounded heel pad than a human."[181]

Doctors Grover Krantz and Jeff Meldrum spent a great deal of time analyzing tracks and the feet presumed to be making them. Although tracks are considered "trace evidence," the trace that tracks leave is due to an actual part of a Sasquatch body and so

---

[178] Krantz 1972b, pages 230-232, 239
[179] Meldrum 2006, page 215
[180] Ibid., page 165
[181] Ibid., page 207

potentially have much to say about structure, locomotion, and even the habits of the animal. Clark quotes John Napier from Napier's 1973 book as saying of the once well-known 17-1/2" Bossburg Cripple tracks, "It is very difficult to conceive of a hoaxer so subtle, so knowledgeable–and so sick–who would deliberately fake a footprint of this nature."[182]

Titcomb makes an important distinction between anecdotal and trace evidence and their helpfulness in establishing anything with certainty about Sasquatch:

> ...the most important thing needed is objective evidence to prove or disprove the sasquatch hypothesis... Anecdotal evidence is that which cannot be verified, confirmed or repeated, regardless of its truth or falsity... The difference between the two types of evidence is the difference between hearing some sort of weird screaming outside a tent in Oregon one night and producing a recorded tape of those noises that can be analyzed in a sound lab by many different people. It is only through evidence that can be analyzed in a scientific fashion, that is, evidence that can be examined in a quantifiable and repeatable way, that a hypothesis can be definitively proven or disproven.[183]

Anecdotes in sufficient numbers and consistency often are more than enough to suggest a testable hypothesis. There is no disputing that the Sasquatch world has plenty of anecdotes. Unfortunately, the record in testing hypotheses has not been impressive.

Meldrum was very specific about what he considered compelling Sasquatch evidence: 1) "indeterminate hair samples that display consistently distinct characteristics"; 2) Eyewitness encounters by "reputable professionals"; 3) "functional analyses" of footprints; 4) Dermatoglyphics [the study of skin ridges on the palms of the hands and fingers and on the soles and toes of the feet]; 5) Recorded vocalizations; and 6) Film that has not been

[182] Clark 1993 page 24
[183] Titcomb 2007

debunked by locomotion experts.[184] Of these six, all but eyewitness encounters (being anecdotal evidence) would be considered *trace evidence.* Unfortunately, since Meldrum published this list in 2006, computer-generated video has become so sophisticated that charges of hoaxed image evidence have become virtually impossible to refute or prove. I would suggest adding to Meldrum's list: 7) Scat; and 8) DNA; though DNA runs into trouble with the lack of a type specimen to compare any new sample against.

Getting back to tracks and feet, Krantz suggested that the Sasquatch ankle, as indicated by tracks and the Patty video, is positioned further forward on the foot than as is seen in *H. sapiens*. He pointed out that this positioning would increase leverage power in walking, helping with the great weight of adults. Meldrum agreed with this forward ankle observation when he noted, "a disproportionate length of the toes and heel by comparison to the human foot."[185] He states elsewhere that shorter [and narrower] heels are associated with lighter hominids.[186] He continues:

> The superficially humanlike appearance is largely the consequence of the inner toe being aligned with the remaining toes, whereas the ape's inner toe diverges much like a thumb… [Yeti tracks also sometimes do appear to have a diverging big toe.] Sasquatch tracks are typically flat with no consistent indication of the true hallmark of the human foot–a fixed longitudinal arch… The sasquatch foot is relatively broader and the sole pad apparently thicker, by comparison to human feet.[187]

Meldrum further describes his findings from research on tracks and the Sasquatch feet that make them:

> In the proposed sasquatch foot with an elongated heel, the

[184] Meldrum 2006, page 275
[185] Ibid., page 165
[186] Ibid., page 247
[187] Ibid., page 223

> gap between the Achilles tendon and the tibia, or shinbone, is wider. When the tendon is comparatively relaxed during the swing phase of the gait, the heel appears to protrude in a nonhuman fashion…. When the calf muscles are contracting and the tendon is taut during the stance phase… the heel no longer seems to protrude and the ankle appears unhumanly thick.[188]

Meldrum again:

> ...the sasquatch foot retains the primitive apelike characteristic of a flat flexible midfoot. The range of motion permitted… allows the heel to function in leverage and propulsion somewhat independently of the relatively prehensile function of the forefoot. In primates this coordinated flexion between the two segments of the foot is referred to as the 'midtarsal break.'[189]

I believe Krantz first postulated the existence in the Sasquatch foot of this "midtarsal break." The tarsals are the seven bones of the ankle and the rear of the foot. The tarsals connect to the more forward bones of the foot, the five *meta*tarsals, which in turn connect to the fourteen phalanges, or toe bones. Simply put, the Sasquatch foot just rear of the midpoint, will flex—the heel rising up relative to the toes. In humans, the foot is relatively inflexible in adults except at the ball. Examination of tracks, track casts, and the Patty film support this additional point of rotation. Page 233 of Meldrum's book has a nice, copyright-protected illustration of a Sasquatch foot pushing off using its midtarsal break next to a less flexible human foot for comparison.

Vendramini's book, with a thesis proposing an extremely adversarial relationship between *H. neanderthalensis* and *H. sapiens*, has a picture of a Neanderthal footprint from a Romanian cave that any Sasquatch investigator would quickly identify as

---

[188] Ibid., page 237
[189] Ibid., pages 234-235

being from one of the Hairy Folk if found in the woods today.[190] Indeed, many Sasquatch witnesses describe something that sounds a lot like a Neanderthal, but generally taller than the Neanderthal maximum that ware aware of, at about six feet. This footprint resemblance may be coincidental or it may be a hint at hybridization between Neanderthal and the Asian version of Sasquatch. Or the track in the Romanian cave may have been assumed to be Neanderthal when it was something else entirely.

On his page 244, Meldrum observes from the Patty film that Patty lifts her toes immediately before setting her foot down. Others observe that Patty swings her toes outward, away from the direction of travel, as she brings her foot forward. It has been suggested that Patty's unique walk is in part derived from dealing with such long feet, much like someone wearing swimming flippers adjusts to walking on land without tripping.

Gregory Forth in his 2022 book, *Between Ape and Human: An Anthropologist on the Trail of a Hidden Hominoid*, points out that the length of the relatively wide feet of *Homo floresiensis* (the popularly dubbed "Hobbit" from very recently dated diminutive remains found on the Indonesian island of Flores), were about 70% of the length of their lower legs compared to only 55% in *H. sapiens*. This extra length was due to long toes and metatarsals (the bones in the foot between the ankle and the visible toes). "...when walking, the very long feet of *Homo floresiensis* would have caused these hominins to bend their knees to a greater extent than modern humans and, in combination with the short legs, to take relatively long strides."[191] Interestingly, much of what Forth describes in *H. floresiensis* anatomy and locomotion is echoed in what is reported for Sasquatch.

Occasionally the best track casts will show the foot equivalent of fingerprints—dermal ridges. In humans, these ridges tend to run across the width of the foot. Jimmy Chilcutt, a crime scene investigator that has fingerprinted many non-human primates, was given full access to Meldrum's track cast collection. What he found was that the dermal ridges that he was able to distinguish

---

[190] Vendramini 2009, page 308

[191] Forth 2022, pages 221-222

were twice as wide as on human tracks and ran the *length* of the tracks unlike on any tracks or feet that he had examined before.[192]

Two last observations on feet. Dr. Henner Fahrenbach, in his 1997-1998 paper *Sasquatch: size, scaling, and statistics*, quoted 10 pounds per square inch (10 psi) as the sole pressure for the unclad human foot. Put a boot on the same foot and the snowshoe effect drops that figure to only 2.5 psi. His examination of Sasquatch casts led to an estimate of 6-9 psi for the unclad Sasquatch foot. The relatively larger flat-footed Sasquatch foot would likely leave *less* deeply imprinted tracks than one of us puny Hairless Ones going barefoot. But put a boot on that Hairless Person and suddenly they are only imprinting about a third as deep as the barefoot Sasquatch.[193]

Coupled with this perhaps surprisingly low *plantar pressure*, the Sasquatch is believed—based on the way it's tracks overlap over solid objects like rocks—to have a sole pad, "probably several inches thick… This thick pad, a tough connective tissue lattice filled with fat… presumably serves to distribute the weight over the sole in a much more even fashion than is the case with humans…"[194] I would also suggest that a sole "lattice filled with fat" would have a relative insulating value against heat or cold.

Skeletal Structure:

Forth suspects that *H. floresiensis* is still alive in remote parts of the large island of Flores. His discussion of the known skeletal remains has some interesting parallels and hints about what Sasquatch skeletal structure might be like. He writes that the remains indicate that *H. floresiensis* was an obligate biped, but that also:

> …other skeletal features suggest floresiensis was a proficient tree climber and… may have spent a good bit of time in trees… Features of floresiensis that facilitate tree climbing

---

192 Meldrum 2006, page 255
193 Fahrenbach 1997-1998
194 Ibid.

> include thick arm bones that were short but, nonetheless, long in relation to the relatively short leg bones… "very curved" hand and foot bones (including those of the toes); and very long feet. …floresiensis' foot lacked a longitudinal arch, meaning that their feet were flat…. The length of the feet, especially, has led specialists to characterize floresiensis as moving bipedally with a "high-stepped gait"…. They must lift each leg high off the ground, thus also bending the leg.[195]

There are parallels between *H. floresiensis* and what we know of Sasquatch skeletal structure and locomotion even though their sizes are vastly different. Perhaps tree-climbing is much more often utilized by Sasquatch than we give them credit for.

ThinkerThunker has done much work on Sasquatch proportions relative to those of humans in his book *Proportional DNA: Solved! Two of Our Greatest Mysteries*. He writes that *H. sapiens* have arms about the same length as their torsos, and that their arms and torsos are both about 30% shorter than their legs. On the other hand, according to ThinkerThunker, Sasquatch arms, legs, and torsos are all about the same length. Proportionally, Sasquatch torsos and arms are both longer than humans, while their legs are proportionally shorter. These proportions were based on Patty.[196]

In another indication that we may be dealing with more than one type or even *species* of creature, the proportions of Patty do not appear to hold throughout the large, hairy hominoid world. Forearms are often described as disproportionally long,[197] beyond what we see in Patty, as is overall arm length.[198]

Height & Weight:

Height and weight also vary widely. In fact, there are some very interesting height outliers. I am going to postpone reviewing those extremes until Volume II because I suspect the very tall

---

[195] Forth 2022, pages 219-222
[196] ThinkerThunker 2023, no page numbers
[197] Isdahl 2021, page 119
[198] Isdahl 2020, pages 237, 238

individuals (over 20' in some cases) are not Sasquatch at all. Hint: I try not to use the otherwise-fitting term "giant" when writing about Sasquatch.

Pyle had a fascinating section based on interviews with "woodsman" Datus Perry who told Pyle that adult male Sasquatch range from eight feet all the way up to eleven feet tall. Grossinger felt that the average North American adult male Sasquatch was 7'-10" (238 cm) tall, but, in perhaps an example of Bergmann's Rule[199] at work, the average height in the Yukon territory was 8'-4" (254 cm).[200]

When it comes to both height and weight, Russel Wiitala in his 2021 book *Sasquatch: Shaman of the Woods*, points out—correctly I believe—that "people have a tendency to consistently underestimate their height… I believe people underestimate their weight, as well. I'm thinking the mature males are all over a thousand pounds…"[201] Fahrenbach agrees regarding the tendency to underestimate height: "in a setting without reference points, observers underestimate the height of a Sasquatch by intuitively assigning anthropomorphic dimensions."[202]

Even so, in weight there may be considerable variation. One of Grossinger's Great White North witnesses described what he saw as about eight feet tall, but skinny.[203]

Odor:

One last aspect of Sasquatch to look at that may or may not be relevant to their physical make-up—the rank odor sometimes associated with them. It would seem that the odor is something they generate as needed or perhaps involuntarily in reaction to

---

199 Bergmann's Rule is the axiom that mammal body size generally increases as average annual temperatures decrease (Bergmann 1847). While Allen suggested in 1877 that "the relative size of exposed portions of the body decreases with decrease of mean temperature." If we were to apply these two rules to Sasquatch, we might expect body size to increase in colder climates, but arm and leg length to decrease.

200 Grossinger 2022, pages 42, 54

201 Wiitala 2021, page 100

202 Fahrenbach 1977-1978

203 Grossinger 2022, page 66

stress. A few witnesses have described brush shelters or caves as containing a powerful lingering stench. Perhaps they generate a distinctive large animal smell like cows or horses. Lapseritis says, "My understanding is that they have a musk-like gland that is activated when they are angry or startled–similar to a skunk's."[204] Male gorillas do have scent glands near their armpits that generate a strong odor when they are excited or alarmed.

---

[204] Lapseritis 2011, page 192

CHAPTER ELEVEN

Sasquatch Vital Statistics & Math

If you do not delight in math, this may not be the chapter for you. You can skip it without losing any of the Patrick story, but I am including this information to complete the picture of just how truly impressive are the Skanicum of the Colville and the Sasquatch of today.

Many researchers and authors have contributed to our knowledge regarding Sasquatch height, weight, and measurements of the foot and hands, etc. A list of the ones that I principally drew upon for the discussion and formulae below would include Duke of *World Bigfoot Radio,* Fahrenbach, Krantz, Meldrum, and Murphy. My apologies to the doubtless scores of authors I did not draw upon. If I had researched this book as thoroughly as the subjects deserve, I would never have written anything at all.

Fahrenbach felt that the width of the heel on any individual Sasquatch lagged proportionally behind length of foot with age. In other words, as the foot grew longer, the heel becomes relatively narrower. He also suggested that heel width had "enormous individual variations" in visible tracks due to the tendency of Sasquatch to lift the heel increasingly with increasing speed, resulting in unnaturally narrow width measurements or none showing at all.[205] In terms of this chapter's topic, we may want to consider Heel Breadth (HB) the least reliable measurement for calculating individual Sasquatch vital statistics.

Jeff Glickman, using a site visit and Patty film frame studies, calculated Patty's height at 7.29' (7' 3-1/2" or 2.22 m) and her foot length as 14.5" (1.21' or 36.9 cm).[206] Fahrenbach points out that this combination yields a "foot-to-height multiplier" of 6.04. My math varied slightly:

$$7.29' \div 1.21' = \mathbf{6.02}$$

Theoretically, **multiply this ratio, 6.02, by the length of any good track in feet to get the height (H) of the maker in feet** (if a Sasquatch). For instance, starting with a 16" track:

16" x 1'/12" = 1.33'
1.33' x **6.02** = 8.03' = 8' 0-3/8" (2.45 m)

Therefore, a 16" track implies an 8' (2.45 m) tall maker.

Duke wields a straight forward way of thinking of this math, handy in the field. **Divide the length of the track in inches by two to get the height in feet**:

$$16'' \div 2 = 8' - 0'' \text{ (2.44 m)}$$

Grover Krantz, who spent a great deal of time thinking about such things and had a doctorate in anthropology to inform his work, suggested another method for determining height (**H** in inches). This method requires three numbers to solve for height. The formula requires the heel breadth as measured below the ankle bone (**HB** in inches). Second, the formula needs the "known" constant for the height to weight relationship for a particular species or body type (I refer to this constant as the Krantz constant, or **K**). Third, it needs the talus bone-bearing area per unit weight, **T**.

[205] Fahrenbach 1977-1978
[206] Glickman personal communication to Fahrenbach 1997

The talus are two articulating surfaces of cartilage [joints] that alone transfer the entire weight of a biped through the two bones of the lower leg, into the foot. The constant **T** is the talus bone-bearing area per unit weight of *any mammal.*

The formula Krantz used to find height from heel breadth is:

$$\mathbf{H} = \mathbf{K} \times \sqrt[3]{(\mathbf{HB}^2 \times \mathbf{T})}$$

HB is measured in the field or from a track cast. Before solving this equation for height, the Krantz constant, **K**, and the talus bone bearing-area per unit weight, **T,** must be known.

The talus supports the two bones of the lower leg—the tibia and the slimmer rearward fibula—with the tibia bearing essentially all of the weight of the animal. It was Krantz's reasoning that bone and cartilage across mammal species have equal weight-bearing capacities. For example, as the weight on the talus doubles, so must the surface area of contact double no matter the individual or species. Quadrupeds are able to spread this weight over two limbs at any one time. In bipeds, one limb must support all of the weight of the organism with each step. There are rumors that Sasquatch have super strong bones, but we must ignore that peculiar possibility until established as fact. The mean of the slightly varying values that Krantz used for the talus bone-bearing area per unit weight in pounds (**W**) is **T** = 24.23 (his **T** varied slightly due to rounding). The square of **HB** in the formula above is a function of area increasing as a square of the increases in linear dimensions since areas are two dimensional properties.

The heel breadth measurement (**HB**) is measured across the widest point of the heel in inches. Using my own foot as an example, that measurement is about 2.6" (6.6 cm).

K, the Krantz constant, the height-to-weight relationship for a particular species or body type is determined by this formula:

$$\mathbf{K} = \mathbf{H} \div \sqrt[3]{\mathbf{W}}$$

The same formula may be re-written to solve for either height or weight:

$$\mathbf{H} = \mathbf{K} \times \sqrt[3]{\mathbf{W}}$$
$$\mathbf{W} = (\mathbf{H}/\mathbf{K})^3$$

The cubed root and third power comes into play because volumes and weights increase to the third power, as linear dimensions change, since volumes exist in three dimensions.

I will use my wispy body as an example. My height, **H**, rounds to 72" (1.83 m). My weight, more or less, 165 pounds (75 kg).

Solving for K, the Krantz constant by body-type:

$$\mathbf{K} = \mathbf{H} \div \sqrt[3]{\mathbf{W}}$$
$$\mathbf{K} = 72 \div \sqrt[3]{165}$$
$$\mathbf{K} = 72 \div 5.48$$
$$\mathbf{K} = 13.1$$

Below, I calculate my own height, using my heel breadth (**HB**) of 2.6" (6.6 cm), a body-type **K** of 13.1, and the universal **T** of 24.23:

$$\mathbf{H} = \mathbf{K} \times \sqrt[3]{(\mathbf{HB}^2 \times \mathbf{T})}$$
$$\mathbf{H} = 13.1 \times \sqrt[3]{(2.6^2 \times 24.23)}$$
$$\mathbf{H} = 13.1 \times \sqrt[3]{(6.76 \times 24.23)}$$
$$\mathbf{H} = 13.1 \times \sqrt[3]{(163.79)}$$
$$\mathbf{H} = 13.1 \times 5.47$$
$$\mathbf{H} = 71.67" = 71\text{-}11/16" = 1.82 \text{ m}$$

71-11/16" (1.82 m) is very close to my declared **H** of 72" (1.83 m). There is some circular reasoning involved with using the Krantz constant, **K**, determined from my height, **H**, to then determine my height through a formula containing **K**, but it does imply that the Talus constant, **T**, is doing its job.

Rather than have to re-calculate **K** values for every biped we encounter, the work completed above for my light frame, along with Krantz's pioneering paper, offer some ready-made values for different body types:

K = 13.1 = Modern male *H. sapiens* light-frame body (equivalent to 6' tall and 165 lbs). This is a useful **K** for a slim human.
K = 12.52 = Modern male H. sapiens body-build (equivalent to 6' tall and 190 lbs). This a useful **K** for a typical medium-build human or that of a young Sasquatch.
K = 10.75 = Body-build equivalent to 6' tall and 300 lbs. This a useful **K** for a lean, but well-muscled or teenage Sasquatch.
K = 9.77 = Body-build equivalent to 6' tall and 400 lbs. This a useful **K** for a heavily-built adult Patty-type of Sasquatch.

In the last chapter I calculated the weight of 700-pound Patty, if she was shrunk to 6'-2", at a whopping *416 pounds*! That aligns very well with the description above of 9.77 **K**. It is very likely that Krantz used Patty for his calculations deriving that particular **K** value.

The equation $\mathbf{H} = \mathbf{K} \times \sqrt[3]{(\mathbf{HB}^2 \times \mathbf{T})}$ may also be solved for the other variables (**T** is a constant 24.23):

$$\mathbf{HB} = \mathbf{H}^3/(\mathbf{T} \times \mathbf{K}^3)$$
$$\mathbf{K} = \mathbf{H}/\sqrt[3]{(\mathbf{HB}^2 \times \mathbf{T})}$$

Put all of this math together and the reference tables produced by Krantz and Grossinger may be expanded:

FOOT LENGTH		HEIGHT			HEEL BREADTH & WEIGHT BY K-VALUE*							
					Light Sasquatch; K = 10.75				Patty-Type Sasquatch; K = 9.77			
					HEEL BREADTH		WEIGHT		HEEL BREADTH		WEIGHT	
Inches	Cm	Feet	Inches	Meters	Inches	Cm	Lbs	Kg	Inches	Cm	Lbs	Kg
6.0	15.2	3.01	36.1	0.92	1.121	2.8	37.93	17.21	1.176	3.0	50.53	22.92
6.5	16.5	3.26	39.1	0.99	1.264	3.2	48.23	21.88	1.326	3.4	64.25	29.14
7.0	17.8	3.51	42.1	1.07	1.412	3.6	60.24	27.32	1.482	3.8	80.24	36.40
7.5	19.1	3.76	45.2	1.15	1.57	4.0	74.09	33.61	1.64	4.2	98.69	44.77
8.0	20.3	4.01	48.2	1.22	1.73	4.4	89.92	40.78	1.81	4.6	120	54
8.5	21.6	4.26	51.2	1.30	1.89	4.8	108	49	1.98	5.0	144	65
9.0	22.9	4.52	54.2	1.38	2.06	5.2	128	58	2.16	5.5	171	77
9.5	24.1	4.77	57.2	1.45	2.23	5.7	151	68	2.34	5.9	201	91
10.0	25.4	5.02	60.2	1.53	2.41	6.1	176	80	2.53	6.4	234	106
10.5	26.7	5.27	63.2	1.61	2.59	6.6	203	92	2.72	6.9	271	123
11.0	27.9	5.52	66.2	1.68	2.78	7.1	234	106	2.92	7.4	311	141
11.5	29.2	5.77	69.2	1.76	2.97	7.6	267	121	3.12	7.9	356	161
12.0	30.5	6.02	72.2	1.83	3.17	8.1	303	138	3.33	8.4	404	183
12.5	31.8	6.27	75.3	1.91	3.37	8.6	343	156	3.54	9.0	457	207
13.0	33.0	6.52	78.3	1.99	3.57	9.1	386	175	3.75	9.5	514	233
13.5	34.3	6.77	81.3	2.06	3.78	9.6	432	196	3.97	10.1	576	261
14.0	35.6	7.02	84.3	2.14	4.00	10.1	482	219	4.19	10.6	642	291
14.5	36.8	7.27	87.3	2.22	4.21	10.7	535	243	4.42	11.2	713	324
15.0	38.1	7.53	90.3	2.29	4.43	11.3	593	269	4.65	11.8	790	358
15.5	39.4	7.78	93.3	2.37	4.65	11.8	654	297	4.88	12.4	871	395
16.0	40.6	8.03	96.3	2.45	4.88	12.4	719	326	5.12	13.0	958	435
16.5	41.9	8.28	99.3	2.52	5.11	13.0	789	358	5.36	13.6	1,051	477
17.0	43.2	8.53	102.3	2.60	5.35	13.6	863	391	5.61	14.2	1,149	521
17.5	44.5	8.78	105.4	2.68	5.58	14.2	941	427	5.86	14.9	1,254	569
18.0	45.7	9.03	108.4	2.75	5.82	14.8	1,024	465	6.11	15.5	1,364	619
18.5	47.0	9.28	111.4	2.83	6.07	15.4	1,112	504	6.37	16.2	1,481	672
19.0	48.3	9.53	114.4	2.91	6.32	16.0	1,205	546	6.63	16.8	1,605	728
19.5	49.5	9.78	117.4	2.98	6.57	16.7	1,302	591	6.89	17.5	1,735	787
20.0	50.8	10.03	120.4	3.06	6.82	17.3	1,405	637	7.16	18.2	1,872	849
20.5	52.1	10.28	123.4	3.13	7.08	18.0	1,513	686	7.43	18.9	2,015	914
21.0	53.3	10.54	126.4	3.21	7.34	18.6	1,626	738	7.70	19.6	2,167	983
21.5	54.6	10.79	129.4	3.29	7.60	19.3	1,745	792	7.98	20.3	2,325	1,055

* K-Value represents the constant height-to-weight relationship for a discrete body type.

T = 24.23 = The universal talus bone bearing-area per unit weight in pounds (used in Heel Breadth calculations)

Note: Krantz 1972b felt that all weights over 1,000 pounds were unlikely to be mechanically possible, yet there are reliable reports of tracks over 18" long.

Table 1: Foot Length, Height, Heel Breadth, & Weight Relationships

CHAPTER TWELVE

Patrick Returns

"I am not overly hopeful of the result of the efforts to civilize the red man. By intermarriage and the heavy death rate, the Indian is doomed to extinction in a few generations. It is a serious grievance to many Indians that the only outlook the Government offers them is that of becoming farmers. Some Indians, like some white people, are natural-born farmers, while others have an inclination to mine and follow other occupations; in fact, few Indians are satisfied with the hum-drum life of the farmer. Many of them take to the wild and reckless life of a cowboy."

--Hal J. Cole; United States Indian Agent;
to the Commissioner of Indian Affairs;
Report of Colville Agency; July 31, 1893

I am pleased to report that the Colville Indians are doing just fine today. Economic development is nurtured by the Colville Tribal Federal Corporation, with revenues of over 100 million dollars a year. They oversee many enterprises including three casinos, a salmon hatchery, convenience stores, and logging operations. The Colville Tribe is a major presence in eastern

Washington and has even purchased traditional lands off the reservation.

More than a century later, tribal enrollment is triple what it was when Patrick was born. And even though unemployment is high and money tight, the twelve tribes are still culturally intact and continue to care for their land.

As we shall see, Patrick did try his hand at farming, but also experienced a "wild and reckless life" as well.

In 1892, the year after Hahissiat's abduction, the northern half of the previously much bigger reservation had reverted back to public domain by an Act of Congress. Then the winter of `92-`93 was so lengthy and severe that many Indian owners of horses and cattle ran out of stored hay and fed their animals the grain seed they had saved for the Spring planting.[207]

I suspect Hahissiat and the very young Patrick returned to Nespelem during this harsh winter about January 1, 1893, when Patrick was approximately six months old. Most likely he traveled up the Columbia by boat for most of the 100-mile return journey. With this suggested timing, his family would have been trying to pass off Patrick as only a three-month old infant.

These were not easy times on the Colville. The reservation had serious challenges before Patrick's arrival. Challenges almost Biblical in proportion, including a plague of locusts:

> Moses' band of Columbias are located about 75 miles from the agency, in the Nespilem [sic] Valley. These Indians, or a

[207] University of Washington Digital Collection: Reports of Agents in Washington; *Report of Colville Agency*; July 31, 1893

> number of them, met with much discouragement last year [the summer of 1890] on account of the ravages done by crickets, and many of them lost their entire crops. The crickets are more numerous this year than last year and have taken almost every crop in the Nespilem Valley, destroying grain, grass, and everything in the vegetable line. I would not be surprised if the government has to render some assistance to the most needy, and especially those who have lost their crops for the past two seasons.[208]

In 1891, there was a treaty negotiation in the town of Nespelem [the modern spelling] in which most of the tribes were convinced to sell the north half of the reservation for $1,500,000 (a dollar an acre). One chief argued that the whites were going to take the land no matter what they did, so the Indians should at least get some money in exchange. Indian residents in the northern half could receive an 80-acre allotment of land and be able to move to the remaining southern half of the reservation (about 1,300,000 acres). Congress didn't authorize the payment until 1906 and the land was then opened to white settlement. The Colville tribes retained hunting and fishing rights in the vacated northern half. "Those living on the ceded portion can keep their homes if they so desire, or they can go to the diminished reservation, which is of ample extent and resource to maintain them in comfort."[209]

Coupled with this land-taking, whites were selling alcohol on the reservation with devastating effect on family dynamics and health:

> ...the whisky traffic, which has been carried on so openly in the past by a class of white scoundrels living near the borders of the reservation, who are utterly devoid of principle or character, and who have robbed, and debased the Indians by selling them whisky, thereby rendering their condition more hopeless... The whisky traffic has caused me more trouble, and

[208]University of Washington Digital Collection: Reports of Agents in Washington. *Report of Colville Agency*; Hal J. Cole, United States Indian Agent; August 15, 1891, page 442

[209] Ibid.

> does more towards retarding civilization than all other evils combined.[210]

Healthcare was in a poor state for years and years among the Colville Indians in general. Hal J. Cole, a sympathetic Indian Agent, wrote in his 1891 report to Washington D.C.:

> It is not to be wondered at that the Indians are decreasing at such an alarming rate. The Indian medicine man is forbidden to practice, and the white physician is located at such a great distance as to be of little or no benefit to these Indians.[211]

The Agency doctor in 1892, E. H. Latham, wrote of his sadness and guilt at the poor health of the Colville Indians:

> I have done all within my power to alleviate the suffering of these poor people. Yet I feel and know that I am not doing them justice.[212]

In 1893 there was no church in the Nespelem valley (Chief Moses was strongly opposed to a Catholic presence[213]), but some of the Moses-Columbia people were going to a Catholic church near Omak lake. The Catholic influence through boarding schools and area churches would eventually have an impact on Patrick and his family. Today the only organized church in Nespelem is the Catholic Sacred Heart Mission.

The need for new roads and improving the ones they already had was growing more obvious, perhaps due to the hardships of the last winter. And the day school in Nespelem burned on Christmas day, 1890.[214] That day school was to have a troubled

---

[210] Ibid.

[211] Ibid.

[212] University of Washington Digital Collection: Reports of Agents in Washington. Report of Physician at Colville Agency; **July 15, 1892**

[213] University of Washington Digital Collection: Reports of Agents in Washington. *Report of Colville Agency*; Hal J. Cole, United States Indian Agent; August 15, 1891, page 445

[214] Ibid.

history in Nespelem, impacting Patrick directly.

On a positive note, Nespelem did have its own sawmill to encourage the construction of frame homes and businesses. The white manager of the sawmill (Mr. Bouska) also had a gristmill to grind grains produced locally by the Indians.[215] Both mills were significant steps in increasing the health and comfort of the local community.

These factors of environment hint at the significant challenges to a happy and healthy life that any child faced in Nespelem or elsewhere on the Colville Reservation. When Patrick arrived, early in 1893, he brought his own personal challenges, but only the lucky and the strong saw their eighteenth birthday on the Colville in the early days of the Reservation.

Nespelem saw and gristmills, ca 1900 by Dr. Edward Latham[216]

[215] Ibid.

[216] University of Washington Digital Collection; Washington State Localities Photographs; https://digitalcollections.lib.washington.edu/digital/collection/wastate/id/656/rec/2

Before Patty

CHAPTER THIRTEEN

The Young Patrick

In this chapter I'll describe what I have found of Patrick's first eighteen years. A good place to start is by revisiting Patrick's *Delayed Birth Registration* from October 30, 1912 (see page 62).[217] Though there are a couple of earlier documents that corroborate Ed Fusch's 1992 paper (which we will also look at in this chapter) this is the one piece of paper that claims to describe Patrick's start on the Turtle Island (North America as termed by some Native Americans).

I believe that when this Certificate of Birth was drafted, Patrick was already ten-years old. There is a typed note on the document: "the information was obtained from the Colville Agency Records and from personal inquiry". The father's name is listed as "Paul [redacted]" [Patrick's family name will be omitted throughout the book to protect living relatives] and it lists Hahissiat Kolockun ("Full maiden name") as the mother. Indian name spellings at this time were often phonetic stabs in the dark by English-speaking transcribers. Patrick's mother's name may have not sounded much like a modern reader would now pronounce these syllables.

[217] 10/30/1912, Patrick [redacted]; Certificate of Birth—Delayed Birth Registration; Division of Vital Statistics; Washington State Department of Health

On this Certificate, Patrick has the same last name as Paul [Pierre Paul was actually his stepfather]. I find that normally Patrick's stepfather's first and middle names are formally given as Pierre Paul, but I have no trouble believing that English speakers favored his middle name and may have only known him by that name. "Pierre" may have been a nod to an early French trapper in the region of Pierre Paul's birth.

Interestingly, this document lists Nespelem as the place of birth. Out of dozens, this is the *only* document that I have found giving Nespelem as Patrick's birthplace. Patrick himself seems to have always given the city of Chelan, Washington that distinction. Remember, the information on this certificate came partly from the "personal inquiry" of a clerk almost ten years after Patrick likely arrived in Nespelem.

This certificate gives his date of birth as 1893 with no month or day. This is another clue that I might be on the right track thinking Hahissiat and Patrick waited until at least January 1, 1893 to arrive in Nespelem. When filling out the form, and making a "personal inquiry" regarding Patrick, people's earliest memories of that small kid would have come from 1893. It might have been natural after all those years to believe that he was born that same year.

Father and mother are listed as "Indian." Apart from the Indian Census, I have not found official documents giving tribal affiliations. Hahissiat/Madeline always appears as within a Moses-Columbia household, though I believe, based on Ed Fusch's paper, that she was married into that tribe, not born of it.

The Certificate was prepared by the Chief Clerk in Nespelem and received and stamped by a "local registrar" on "Oct 30 1912".

Patrick may have had some local fame in 1912 based on his odd looks, though I do not believe that he looked as radically peculiar as local lore had him when Ed was doing his research in the late 1980s. It occurs to me also that Ed Fusch may not have been an impartial collector of information on Skanicum and on Patrick. He likely had a firm concept of what a Sasquatch or a Sasquatch/Human hybrid should look like before he began his ethnographic interviews on the Colville. This predisposition might have colored the questions he asked and what he recorded from answers and testimony (a predisposition that others, like me, might share in retelling the story). In the two pictures of Patrick that I have found (and will share!), he looks odd, but I do not see some of what Ed Fusch transcribed from informant testimony: "...very large lower jaw, a very large wide mouth with straight upper and lower lips.... He was kind of stooped, or humpbacked. His ears were elongated upwards (peaked) and bent outward at the top. ...very ugly..."[218]

When I look at the pictures of the older version of Patrick, I do not see this list of supposed defects, but I do see someone that looks odd. Patrick the man looked *funny* and that must have been the case for Patrick the boy as well. As he grew, his appearance may have been a cause of discomfort and concern for his mother, Madeline Hahissiat, and his adoptive father, Pierre Paul. Perhaps they hoped he would grow out of some of his awkwardness, and perhaps he did. Maybe they kept him largely out of view and he didn't play much with other children or attend large gatherings. His parents may have been unable to shake discomfort over the looks of their child—the oddities more obvious to them than to those more distant from Patrick.

One benefit Patrick seems to have enjoyed is that of hybrid vigor. His was a remarkably long life, despite the risk factors he traversed. Perhaps a youth mostly secluded from other children, though lonely, did shield him from some childhood diseases, including the deadly tuberculosis.

Likely he was the victim of teasing over his looks, his small size, and perhaps the persistent rumors about whom his father really was. I suspect Patrick had to learn how to conduct himself in a fight and if Ed's description of him as being smart, having very long arms, and large hands are true, these qualities may have helped to counterbalance his size.

Despite all the strikes against Patrick from birth, the Indian women that Fusch interviewed said that Patrick was "a gentle man."[219] This description also is not well-supported by the details of Patrick's life that I have uncovered.

---

[218] Fusch 1992a, page 21 and Burtsev 2015, page 306

[219] Fusch 1992a, page 21

In the year before August 1892 [the year in which I believe Patrick was born], the tribal court convicted 5 cases of "wife beating," 41 of drunkenness, 3 of larceny, 3 of robbery, and 6 of adultery. Many on the Colville still did not see the need for the white man's paper before living together, so I don't assign too much meaning to the adultery cases, but alcohol probably was much, much more of a problem than the number of convictions would indicate and likely had something to do with the wife beating cases as well. The social fabric of the twelve tribes was stretched and torn during this transition time from the freedom before the European invasion, to this new and unhappy period of confinement on the reservation. Many of the former adult roles were now gone or shifting in directions perhaps not particularly valued in the context of their old lives.

Ed Fush's paper reported of Patrick that "He attended school on the reservation, was 'very smart'…"[220] I have no doubt that he was smart. The 1920 United States Federal Census indicates that he was able to read and to write—and the four Patrick signatures that I have found on documents spanning 37 years all appear to have been quite legibly written by the same hand (presumably Patrick's!)—but the 1940 United States Federal Census indicates that the most schooling that he was able to complete was the second grade. His intelligence would seem to be one that he was born with, not one gained from schooling. For better or for worse, in Patrick's time less than half of school age Indian children were able to attend the two reservation boarding schools and the frequently closed day schools.[221]

For Patrick's early life he lived in a nuclear family composed only of his stepfather, Pierre Paul, and his mother Hahissiat (now officially going by the English name, Madeline). It was a few years before Madeline would have another child—a half-sister to Patrick.

Patrick's First Year (Born About June 1, 1892)

---

[220] Fusch 1992a, page 21

[221] University of Washington Digital Collection: Reports of Agents in Washington. *Report of Colville Agency*; Hal J. Cole, United States Indian Agent, July 31, 1893, page 325

Tuberculosis was the great killer of children on the Colville. Present were both the better known form that attacks the lungs and internal organs (commonly known in Patrick's time as *consumption*) and the less deadly form called scrofula, which attacks the lymph nodes and manifests as disfiguring lesions on the neck. From C. K. Smith, Physician on the Coeur d'Alene Reservation in 1893:

> The great plague of this tribe, and in fact of all Indian tribes, is tuberculosis, generally of the lungs. The majority of the deaths are from this cause. Probably the principal factors in producing this disease is the change in diet and the change from outdoor life and occupations to the more confining and sedentary habits of those in the schools and those living in closed dwellings. It is much more prevalent amongst children than grown persons.[222]

The infant Patrick was to escape tuberculosis (the White Plague), but the disease was to have a huge impact throughout Patrick's life. In the short term though, the winter of 1892/1893—in which the infant Patrick may have returned to the Reservation—was unexpectedly harsh:

> The Indians [sic] loss in stock during the past winter amounted to several thousand head, principally horses, although some lost heavily in cattle. The majority of Indians put up feed for their stock, but not a sufficient amount, however, to feed them during the long and severe winter.[223]

Patrick One-Year-Old About June 1, 1893

In the 1894 Colville Agency report (reports covered June 1st of the previous calendar year through May 31st of the year of the report) the Indian Agent wrote:

---

[222] Ibid.,, page 325
[223] Ibid.

> Most all of the crimes committed can be traced to whisky, and as long as Indians can obtain it I suppose crimes of a serious nature will be committed. Quite a number of persons have been convicted for selling liquor to Indians, principally through the efforts of the U.S. Marshall and his deputies. Others introducing liquor on the reservation have been run off and their stock destroyed.[224]

Neither Nez Percé or Moses-Columbia men joined the tribal police. "Up to the present time Chief Moses, of the Columbias, and Joseph, of the Nez Percés, refuse to supply any of the police or have any of their number on such duty."[225] I suspect there were not many tribal police in Nespelem as a consequence and that Patrick's people were generally on their own to work out disputes and deal with the effects of alcohol in their community.

Agency police were not all that was avoided in Nespelem. In the 1894 Report, Captain Bubb writes that "As a rule the Indians dislike sending their children away to school, even so short a distance as the agency boarding school…" Patrick's failure to attend past second grade was probably more a reasonable decision of his parents to shield him from disease, harsh discipline, and cultural destruction than as a way to keep him out of the public eye or away from the teasing of his peers. As we will find out later, chronic closures of the local day school may have been the biggest limiting factor for Patrick's education.

The love of horses seems to have been universal among the twelve tribes and Patrick probably learned how to ride at a very young age. It is likely that during the time of Patrick's youth, his stepfather was beginning to try his reluctant hand at farming, and anything else that might earn some money. Farming was not a natural fit for the Nespelem valley, but there was pressure and rewards from the Washington D.C. bureaucrats to make the best of the agrarian life. From the 1894 Colville Agency report:

---

[224] University of Washington Digital Collection: Reports of Agents in Washington. *Report of Colville Agency*; August 21, 1894, Captain U.S. Army, Acting U.S. Indian Agent, Jno. W. Bubb, pages 312-313
[225] Ibid.

> The Colville and Spokane reservations are not generally adapted to farming, being mostly timbered, mountainous, and rocky. The Indians of the various tribes carry on farming and utilize pretty much all of the available land susceptible of cultivation and irrigation. They raise wheat, oats, barley, and all kinds of vegetables, nearly sufficient for maintenance of their families. …Farms, as rule are small… All have some stock, principally horses. I have endeavored to impress them with the importance of increasing their cattle herds, thus utilizing the large amount of valuable grazing lands they possess. Outside of farming and stock raising many engage in freighting, both for the Government and private parties; others pick hops, fish, and hunt.[226]

Patrick Two-Years-Old About June 1, 1894

Crickets and long winters were behind the Nespelem residents for now and hunger was no longer such a concern for the local farmers:

> The past year was a more favorable one so far as crops are concerned, the yield being all that could be expected, but the low price of grain, particularly wheat, scarcity of work, etc., has told on the Indian as well as the white man.[227]

My parents—in Europe during the Great Depression of the 1930s—told me about their experience of those times: "We were poor before the Depression, poor during the Depression, and poor after the Depression. We hardly noticed." Perhaps the Panic of 1893 had the same effect on the Colville, but for five years the United States as a whole was subject to high unemployment, bank failures, strikes, and marches. Some of that must have trickled down to the people of the Colville, struggling as they were make

[226] Ibid.,, pages 311-312

[227] University of Washington Digital Collection: Reports of Agents in Washington. *Report of Colville Agency*; August 16, 1895, Captain U.S. Army, Acting U.S. Indian Agent, Jno. W. Bubb, page 313

it through each year.

The growing trend on the Reservation, and the obvious choice for the men (and women) who loved their horses was to raise cattle:

> I have been looking over this country pretty thoroughly and see no prospect of their ever extending their farms very much. There is plenty of valuable grazing land, however, and it would seem to me to be the better policy to try and induce them to engage more largely in cattle raising. Like most other Indians, they are wedded to the pony, and think if they have a large pony herd they are rich. I notice that all those who have obtained a start in cattle are much better off. Of the cattle issued to the Columbias some years ago there is hardly a hoof left.[228]

As you may remember, the Nespelem day school suffered a fire on December 25, 1890, but sometime in this year of June 1, 1893 to May 31, 1894 the school was rebuilt along with a teacherage. The Indian Agent hoped to steer education on the Reservation in the direction that he saw most suitable:

> The Indian now needs the practical education which will more quickly fit him for self-support and future citizenship. It seems to me that the results so far attained in the matter of education are not equivalent to the expenditure. The number of young men and women who can read and write is very small and I attribute this largely to the fact that the schools connected to this agency have not progressed with the times.[229]

An extremely discouraging statistic these Agency Reports reveal is of consistently decreasing population numbers across the twelve tribes. Poverty, addiction, and disease more than offset any advantage supposedly awarded by a more "civilized" and

---

[228] Ibid.
[229] Ibid.

sedentary life.

The white world kept pressing in on the reservation. Bubb, writing of the Indian police force, "They have been kept busy constantly keeping off prospectors and trespassers, in addition to their usual duties among their tribes."[230]

Patrick Three -Years-Old About June 1, 1895

Under a new Indian Agent, the Nespelem Day School continued to have its problems:

> Day school No. 2, located at Nespilum [sic], 60 miles from the agency on the south half of the Colville Reservation, was opened the 1st of February this year [1896]... This school has been practically a failure from the beginning by reason of having the opposition of the two head chiefs in that locality, Moses, of the Columbias, and Joseph, of Joseph's band of Nez Percés. It is believed, however, that with another year they will give it their support and encouragement, and that satisfactory progress will have been shown in the matter of education among their people. ... The total enrollment at this school was 17... capacity, 50...[231]

Joseph's band also frustrated the new Agent:

> Joseph's band of Nez Percés are still living in their teepees, although many of them have good homes. They do but little farming, and depend almost wholly upon the issue of government rations."[232]

Although the Agent grudgingly, and with insult, admitted that times and available resources made any gains extremely challenging:

[230] Ibid

[231] University of Washington Digital Collection: Reports of Agents in Washington. *Report of Colville Agency*; August 22, 1896; Geo. H. Newman, United States Indian Agent, page 310

[232] Ibid., page 311

> Besides farming and stock raising, there is very little other employment for them. Some few engage in freighting, while others fish, hunt, and pick hops… Many of them, however, who have good farms are shiftless and lazy and seem content to raise barely sufficient to keep them from starvation. It is very difficult to make them see the necessity of improving their farms and cultivating the land so that it will yield larger returns per acre. Most of them still have bands of ponies… They are accumulating cattle slowly.[233]

The same Report acknowledges that the efforts to get the Colville Indians to accept an allotment of their own private land had gone nowhere. The concept of private land was unnatural to a people who considered that land was to be used in common by the tribe:

> The situation, however, would be much simplified if it could be assured they would be permitted to live on their allotments peacefully, and not be taken advantage of by their white neighbors in all sorts of ways. The Indian is considered legitimate prey by a certain class of people who will not hesitate to employ any means to drive him from his allotment, and force him, if possible, to abandon it. These people religiously think an Indian has no rights a white man is bound to respect.[234]

The seeking of visions through alcohol continued to plague the Colville:

> Whisky is unquestionably the bane of the Indian. It is very difficult under the laws to obtain sufficient evidence to convict men guilty of selling it to them. Especially is this so on so large a reservation as the Colville, which is surrounded by quite a number of small towns, where it can always be

[233] Ibid., page 311
[234] Ibid., page 311

easily procured.[235]

And the illegal miners continued to spill over the Reservation boundary in Patrick's fourth year:

> ...prospectors have swarmed all over both reservations during the past few months and it has been extremely difficult to get them off. Several hundred notices of mineral location, with the stakes put up to mark the claims, have been destroyed by the Indian police.[236]

Against these kinds of attitudes and direct harm from their white neighbors, the Agent thought the white man's religion could offer some form of compensation. "I think there is a splendid field for missionary work among these Indians. A church should be established at the Nespilem [sic] station..."[237]

Patrick Four-Years-Old About June 1, 1896

In his 1897 Report, the same Agent give us an impression of excellent Skanicum habitat: "The country comprising the Colville and Spokane reservations is rough and mountainous in character..."[238] Although the trespassing prospectors may have eventually explored every remote valley and rocky crag.

The outlook for the Colville Indians continued poor even according to the Indian Agent responsible for their welfare. These next two excerpts are long, but they extend the observations of someone that was there at the time, seeing the odds against these peoples who had already been dealing with the incremental loss of their lifestyles and resources for *decades*. In this excerpt, Agent Newman describes another season of crop failures (the summer of 1896) and again the need to depend more heavily on raising cattle:

---

[235] Ibid., page 312

[236] Ibid., page 313

[237] Ibid., page 312

[238] University of Washington Digital Collection: Reports of Agents in Washington. *Report of Colville Agency*; August 15, 1897; Geo. H. Newman, United States Indian Agent, page 288

> The almost entire failure of crops last year, and the lack of means to purchase seed for this year's planting, and the gloomy outlook for the approaching winter, are enough to discourage... Farming is the principal occupation of the Indians in charge of this agency. In fact, there is very little other employment for them... The Indians of the Colville and Spokane reservations, as a rule, are very poor, and in my opinion will never become self-supporting if they shall have to depend exclusively on farming for a living. As both reservations are a great deal better adapted for stock raising—especially cattle—than agricultural products, every effort should be made to encourage them along that line.[239]

He continues to point out the environmental and social challenges his charges were up against:

> When I look around and note the rough mountainous character of the country and see the gradual encroachment of the whites upon all sides, the scarcity of game, the almost utter lack of employment, the majority of them with only a few acres of ground to cultivate, and dependent almost entirely on the few bushels of grain they raise, I am astonished at the progress they have made, while at the same time I wonder how they have managed to live.[240]

The Nespelem Day School was in trouble again. In the 1896 to 1897 school year it still struggled with low attendance and closures because of staffing problems. It finally reopened on April 1, 1897 and had three pretty good months.

This next Agency Report excerpt that I include is an example of the nature of those times before the turn into the twentieth century:

> On the 4th of June last a Chinaman was shot and killed

[239] Ibid., page 289

[240] Ibid., pages 289-290

by a Nez Percé Indian, a member of Joseph's band. The shooting occurred just below Nespilem [sic], on the Columbia river. The Indian claims the shooting was accidental—that he fired at some wild ducks on the river, not seeing the Chinaman on the opposite side, and did not know until several days afterwards that the shot had taken effect. I investigated the matter carefully and am convinced that this is the true version of the affair, but I have notified the United States marshal...[241]

Patrick Five-Years-Old About June 1, 1897

The Reservation was opened to mining in 1898, when I believe Patrick was five-years old. This brought miners in great numbers to Nespelem and the surrounding areas and increased the sense of being overrun by the white world with its liquor, illness, and crime. The Colville had already endured waves of trappers, farmers, and lumbermen. The Moses-Columbia traditional territory had once been 4.3 million expansive acres, but now many Moses-Columbia had been moved north to the Reservation and into less hospitable areas like the Nespelem valley, though that area was still close to the Columbia.

There was a silver lining to the influx of miners that will soon show up in Agent Reports.

This year of Patrick's life brought yet another new Indian Agent, Albert M. Anderson. Mr. Anderson had in common his predecessors' frustration with the Nez Percé: "...stringent rules must be applied and enforced with vigor."[242]

The previous summer and fall had brought good harvests and, slowly, it would seem that the shift to raising cattle had been occurring in this last decade of the 1800s. Hop-picking continued to be a source of cash income for some of the band. Later, we will see evidence that Patrick himself would pick hops to support himself and his dependents.

---

[241] Ibid., page 291

[242] University of Washington Digital Collection: Reports of Agents in Washington. *Report of Colville Agency*; August 20, 1898; Albert M. Anderson, United States Indian Agent, page 298

> Moses and his band of Columbias… are as a class industrious and thrifty, and a decided improvement is manifest in the condition of these Indians in the past year. …their crops have received more attention than heretofore, and their harvest is abundant. …Many of them are curtailing their traditional bands of ponies and entering into the more profitable employment of raising cattle… They make annual journeys to the hop fields, and by their labor earn enough money to purchase clothing and many necessaries for their households.[243]

The Nespelem Day School succeeded in staying open for nine months in the school year of 1896/1897. Patrick was on the cusp of being old enough to attend, but he appeared on paper to be at least three months *younger* than his actual parentage would imply. I speculate that he did not attend during the school year of 1897/1898.

No doubt against the wishes of Chief Moses (personally prosperous and generous with a farm of over 1,000 acres and cattle and horses)[244], the Catholic religion was slowly gaining ground in Nespelem and amongst his Columbia band, a sign of progress to the white world:

> The morality of this tribe [Moses-Columbia], while not good, is at least encouraging. Living as they do, away from the influence of any church, their religious training is easily forgotten and they quickly lapse into their original and more primitive customs and practices. A number of them are members of the Catholic Church and attend divine services at St. Mary's Mission two or three times a year. A number of marriage ceremonies have been performed and quite a number of their children baptized.[245]

Patrick Six-Years-Old About June 1, 1898

---

243 Ibid., pages 298-299
244 Ibid., page 299
245 Ibid., page 299

The end of an era came in Patrick's seventh year. In an Agency Report passage that reflects sincere admiration, Anderson writes:

> The Columbia tribe have sustained a severe loss during the past year in the death of their chief, Moses. ...he succumbed to the inevitable March 25, 1899. ...Moses came of fighting stock and was a chief by inheritance. His ancestors for generations had been chiefs and possessed undaunted courage and bravery. ...He often contested for supremacy with white people, backing up the merits of his horse with cash. His judgement was rarely misplaced. ...Though stern in governing, he was kind to the weak and generous to those in want, caring for the old and indigent, supplying them with provisions from his own larder, and often gave them money to purchase the necessaries of life.[246]

Chief Moses was a true champion of his people and he is remembered still with respect on the Colville. His death was probably a blow to the Colville people, shaking their confidence in being able to meet the continuous challenges of these unstable new times.

But there was the silver-lining of capitalism to the cloud of miner's camps that I promised you earlier:

> The extension of the mineral laws to the Colville Reservation, while having its injurious effects and drawbacks, has its redeeming feature. Prospectors and mining camps create a market for the produce of the Indians, and, as a general rule, the prices are far more in advance of those in a more open market outside.[247]

Patrick Seven-Years-Old About June 1, 1899

---

[246] University of Washington Digital Collection: Reports of Agents in Washington. *Report of Colville Agency*; August 25, 1899; Albert M. Anderson, United States Indian Agent, page 354

[247] Ibid., page 354

This fiscal year of the new Agency Report (June 1, 1899 to May 31, 1900) had both good news and bad news for Patrick, with the good news perhaps *finally* outweighing the bad news. The Colville Indians had quickly made sweet lemonade out of the bitter lemons of the whites on their reservation:

> …I can state without the slightest hesitation that there has been a decided improvement in agriculture and all branches of industry in which they are engaged. The past year has been quite favorable in the region occupied by these Indians, which has added a stimulus to agriculture and stock raising. The past winter was an exceptionally mild one, and cattle and horses subsisted on the ranges without being fed or sheltered. Very few of them died, and a large amount of fodder has been carried over for another year. Large tracts of land on benches of higher elevation have been inclosed [sic] by Indians during the year by substantial fences and sown in wheat and oats. …A few of them grow a small amount of timothy [*Phleum pratense*: a European perennial grass grown for hay], for which they find ready sale in the different mining camps throughout the reservation, and they are generally paid a better price for their produce than that obtainable in the more open market. Those markets are a great benefit to the Indians and are a source of revenue throughout the year, and finding ready sale for their produce has stimulated them to enlarge their areas of cultivation and pay more attention to the care of their farms and gardens.[248]

It must have been very satisfying to take from the whites for a change.

I won't repeat the intemperate details that he included at length, but this one sentence from Mr. Anderson's Report sums up how he was feeling about the Nez Percé: "I have not the slightest hesitation in stating that Chief Joseph and his people are

---

[248] University of Washington Digital Collection: Reports of Agents in Washington. *Report of Colville Agency*; September 29, 1900; Albert M. Anderson, United States Indian Agent, page 392

a nonprogressive and indolent set."[249]

Unfortunately, times weren't all about up-charging the miners and driving the Indian Agent nuts. An old disease made its reappearance right where Patrick was living:

> During the past winter smallpox broke out among the Indians residing on the south half of the Colville Reservation. On November 25, 1899, the agency physician reported that the disease existed among the Columbia tribe, residing in the vicinity of Nespilem [sic] subagency. ...In order to contend with so grave an emergency, it was considered advisable and essential to establish and maintain a general quarantine, which went into effect on December 5. Indians were forbidden to leave their homes... The infected district comprised a territory about 15 miles long and about 6 miles wide, and covered a very rough and abrupt country. ...In all, there were 57 cases with 6 fatalities. Those who succumbed to the disease were old people... The quarantine was raised on March 25...[250]

Presumably, if Patrick had started school this year, he was forced to stop due to the quarantine after only about three months.

This was not the first time that the Moses-Columbia had weathered a smallpox outbreak. Between 1769 and 1780, smallpox tore through the upper Columbia drainages, causing horrific death rates among the Colville and the Nez Percé. Two different European expeditions to the Puget Sound and the lower Columbia areas found survivors in 1792 horribly pitted with pox scars and suffering much misery from blindness. Entire villages were found abandoned, the structures returning to the earth.[251] The descendants in the 1890s of the devastating Washington state outbreaks of the 1770s were less vulnerable to the disease than their ancestors after the horrendous winnowing of those carrying the least protective genetics.

Patrick Eight-Years-Old About June 1, 1900

[249] Ibid., page 393

[250] Ibid., Pages 392-393

[251] Lange 2003

Smallpox returned to the Colville again in the winter of 1900-1901 in Patrick's ninth year, with a greater spread across the reservation. Fortunately, many lessons had been learned during the initial outbreak at Nespelem and the Agency employees and public health officials were more prepared to react. 252 cases were recorded with 26 fatalities. Presumably, Nespelem had already experienced the worst during the previous winter and was left minimally impacted.[252]

Nespelem at or shortly before 1900 by Dr. Edward Latham. Mercantile on left, Hotel in center. [253]

Unfortunately, after two-years of quarantine interruptions, the Nespelem Day School was closed at the end of this school year, "...having hardly justified continued maintenance."[254] The Nespelem School was not to be re-opened until the school year of 1906/1907 when Patrick was already fourteen years old and probably then crucial to Pierre Paul's farming and ranching operation. A document from later in Patrick's life says that he only attended school up to grade 2. His attendance may have come during the school years of 1898-1899, and the two smallpox-

[252] University of Washington Digital Collection: Reports of Agents in Washington. *Report of Colville Agency*; September 12, 1901; Albert M. Anderson, United States Indian Agent, page 385

[253] University of Washington Digital Collection; Washington State Localities Photographs; https://digitalcollections.lib.washington.edu/digital/collection/wastate/id/96/rec/3

[254] University of Washington Digital Collection: Reports of Agents in Washington. *Report of Colville Agency*; September 12, 1901; Albert M. Anderson, United States Indian Agent, page 385

quarantine interrupted years of 1899-1900, and 1900-1901. For a person described as "highly intelligent," his lack of educational resources may have had a profound negative impact on the rest of his life.

Hotel in Nespelem, ca 1900 by Dr. Edward Latham
[255]

Indian Agent Anderson felt that he was making headway with the Nez Percé, though it appears he was doing so mostly through the deliberate threat of starvation:

> I am glad to report, too, that Chief Joseph's band of the Nez Percés has done more work and has been less troublesome than ever before, this, in my opinion, having been a direct and logical result of reducing rations—a process which I am inclined to push further.[256]

The good news/bad news effects of the opening of Patrick's south half of the Colville Reservation to mineral staking continued to play out regarding eager buyers of farm and ranch products versus the availability of alcohol:

---

[255] University of Washington Digital Collection; Washington State Localities Photographs; https://digitalcollections.lib.washington.edu/digital/collection/wastate/id/336/rec/1

[256] Ibid., page 385

> The influx of white population, which has resulted from that action, has been of benefit to the Indians in creating markets for their produce, but it has also made it extremely difficult to prevent the sale of intoxicants to them, and has resulted in much aggression upon their rights in various ways.[257]

In hindsight, allotment proved to be devastating to the socio-economic organization of most tribes in the United States. The coming land allotment process for Colville was sounding alarm bells even for this rather harsh Indian Agent:

> As to the liquor traffic, it would seem that absolute defeat is at hand in so far as allotted Indians are concerned. Judge Hanford, of the United Sates circuit court for the District of Washington, recently ruled that an allotted Indian is a full-fledged American citizen and at entire liberty to buy intoxicants when and where he pleases. This may be a correct interpretation of existing law, but it is a very unfortunate decision for the allotted Indian.[258]

I apologize for how often I bring up the harms of alcohol for the Colville Indians, however, as you will read later, alcohol will have a tremendous impact on Patrick's life.

Patrick Nine-Years-Old About June 1, 1901

I did not find anything specific to report about Patrick's tenth year. Indian Agent Anderson did reaffirm that the Colville still contained remote lands in which large hairy bipeds might find concealment: "The territory under the jurisdiction of this agency is very large and much of it extremely inaccessible."[259]

---

[257] Ibid ., page 384
[258] Ibid., page 384
[259] University of Washington Digital Collection: Reports of Agents in Washington. *Report of Colville Agency*; September 2, 1902; Albert M. Anderson, United States Indian Agent, page 354

Patrick Ten-Years-Old About June 1, 1902

If there was a tendency to think of North America's First Nations peoples as needing government assistance to survive, apparently this was not the case on the Colville in the early 1900s according to Agent Anderson. "Only a few old and indigent of the Spokans [sic] and of the various Colville tribes receive any material assistance from the Government, and none receive any annuities." However, alcohol continued to create problems for the Indians, with no end in sight: "…the whisky traffic, particularly on the south half of the Colville Reservation has continued to give much trouble, and has had the usual effects on the welfare and progress of the Indians where I have been unable to stop it."[260]

Patrick Eleven-Years-Old About June 1, 1903

Despite diminished crops due to drought in 1903, there was new income coming from grazing leases on Reservation land, though this income may have gone right into the United States Treasury instead of being held in trust for the Colville tribes: "At first opposed by the Indians when first inaugurated a year or two ago, they have now ceased their opposition to the grazing of white men's cattle on the more or less extensive ranges of their portion of the reservation…" Special United States Indian Agent in Charge, S. L. Taggart (Indian Agent Anderson had been replaced without explanation), reported grazing income of $5,482.73 (over $200,000 in 2025's inflated dollars).[261]

Patrick Twelve-Years-Old About June 1, 1904

The death at Nespilem [sic], on the 21st of September, 1904,

---

[260] University of Washington Digital Collection: Reports of Agents in Washington. *Report of Colville Agency*; September 4, 1903; Albert M. Anderson, United States Indian Agent, pages 331-332

[261] University of Washington Digital Collection: Reports of Agents in Washington. *Report of Colville Agency*; July 31, 1904; S. L. Taggart, Special United States Indian Agent in Charge, pages 349-350

> of Joseph, chief of the Nez Percés, was a noteworthy event. Whatever the faults of this man, he was a born leader and the peer of our historically great red men. On June 20, 1905, at Nespilem, the place of his exile, a handsome monument of white marble was dedicated to his memory, with appropriate ceremonies...[262]

Though there still was one old and blind war chief of the Nez Percé still alive at Nespelem, the death of Chief Joseph seemed to bring to an end the era when the young were visibly and daily reminded of a past that seemed, in retrospect, more glorious than the struggles of the present times.

In this year the Agent affirmed that, "The Colville Indians... are self-supporting, and serious cases of destitution are seldom known."[263]

For some time I attempted to track annual changes in tribal populations during Patrick's early years until I read this passage in the 1905 report:

> Great care has been used in taking the census, which is nearly accurate and entirely free from estimates. A material reduction of population is shown. ... The number of children of school age has heretofore probably been overestimated, while tuberculosis and hereditary scrofula have made serious ravages among the youngsters, an unusually large proportion of them being thereby debarred from school privileges.[264]

It turns out that earlier reports, relying on estimates from local white farmers, often double-counted families. Soon a very exacting annual Indian Census would begin, but when Patrick was twelve the trend was alarming: The Indian population was dropping every year. In 1905, on the entire Colville Reservation, there was only about 279 men, women, and children that

---

[262] University of Washington Digital Collection: Reports of Agents in Washington. *Report of Colville Agency*; August 29, 1905; Jno. McA. Webster, Captain, U.S. Army, Indian Agent, page 357

[263] Ibid., page 355

[264] Ibid., pages 356-357

identified as Moses-Columbia. Even more alarming, the Joseph's band of the Nez Percé was down to 89 people.

Whites selling alcohol to Indians played a big part in this declining population, but it was proving impossible to stop this unconscionable trade as Agent Webster pointed out: "...juries in this country refusing to convict on Indian testimony alone..."[265]

Webster gives us a glimpse of the implacable white invasion with which the Indians (and the hidden Stick Indians) were coping:

> The opening of the Colville Reservation to mineral entry brought a train of evils in its wake. Towns have been built on placer mining claims, and all sorts of businesses, excepting that of mining, are carried on. Saloons were established in the heart of the reservation...[266]

This year, or the next, Patrick would see the birth of his half-sister, Christine. There was a large gap between Patrick's birth, in 1892, and Christine's birth now. There may have been other unrecorded children in the meantime that died young.

Patrick had a kid sister now and perhaps he enjoyed this semblance of a nuclear family, even though his responsibilities probably grew.

Patrick Thirteen-Years-Old About June 1, 1905

Wanting to give Washington D.C. something more positive to consider, Agent Webster ended his 1906 report with this good news:

> ...the certainty of abundant crops this season insures to them a better financial condition than they have ever before known. While progress has been slow the outlook is encouraging, and I have faith in a better future for these red

[265] Ibid., page 356
[266] Ibid., page 356

men.[267]

Change was coming to the entire country and even finding its way to the Colville:

> Means of transportation between the various districts include railroads and river steamers around the outer boundaries; with a few good, some indifferent, and many bad wagon roads and trails in the interior; the Colville and Spokane reservations having no means of rapid transit. This deficiency will be remedied in the near future if some of the proposed steam and electric lines are built, and much of this there can be no doubt, for the country is rich in productive land, much of which is already being developed…[268]

Unfortunately, access for the Colville Indians also meant access for the white world, eager to profit from Reservation resources, even if that profit required theft. The Indian Agents charged to manage the Reservation lands were unable to stop the stealing:

> The rich soil of the valleys of the Colville Reservation along the Columbia, San Poil, Nespilem, and Okanogan rivers has tempted the cupidity of numerous persons who have made thousands of entries of quartz and placer mining claims on splendid fruit, wheat, and timber lands. In this process many Indians have been cheated, frightened, cajoled, or bought out of their rightful possessions… White homesteaders have diverted most, and in some cases all, of the water from sources used by the Indians for many years, even their drinking-water supply in some instances being completely cut off.[269]

This Agency Report didn't fail to mention alcohol sales: "The

---

[267] University of Washington Digital Collection: Reports of Agents in Washington. *Report of Colville Agency*; August 17, 1906; Jno. McA. Webster, Captain, U.S. Army, Indian Agent, page 372

[268] Ibid., page 371

[269] Ibid., page 371-372

local civil authorities seem indifferent or powerless and the whisky element holds the upper hand."[270]

Patrick Fourteen-Years-Old About June 1, 1906

No 1907 agency report is available, but the 1906 report promised the Nespelem Day School would reopen for the 1906-1907 school year when Patrick would have been fourteen... too late for the intelligent and resourceful Patrick—his formal schooling was behind him now.

Patrick Fifteen-Years-Old About June 1, 1907

A 1907 document described the process of the ceding of the north half of the Colville Reservation and the process of allotting land to the Indians on the south half—the Colville Reservation as we know it today:

> The diminished reservation is estimated to contain 1,300,000 acres. Under the act of March 22, 1906 (34 Stat. L., 80), these lands are to be surveyed and allotted... and the surplus lands are to be opened for settlement and entry under the homestead laws of the United States. ...The work of allotting will be taken up during the present fiscal year and will be well under way by the summer of 1909. There are about 2,452 Indians to be allotted.[271]

Patrick Sixteen-Years-Old About June 1, 1908

I was unable to find a Colville Agency Report or anything else directly relevant to Patrick's life regarding his seventeenth year.

Patrick Seventeen-Years-Old About June 1, 1909

This is the earliest year in which I have located documentation

---

[270] Ibid., page 371

[271] University of Washington Digital Collection: Report of Commissioner of Indian Affairs; Field Work on Reservations; 1908, page 61

of Patrick individually (besides his Delayed Birth Certificate)—in the June 30, 1909 Indian Census. Confusingly, it gives his age as 12, but in the next Census year (1910) his age is recorded closer to correct, but still incorrect at 15 years old. However, this 1909 Indian Census for the Colville Agency gives his father as Pierre Paul and his mother as Madeline, confirming his identification. It also shows him with an unnamed three-year old sister, which lines up well with later information about his half-sister, Christine, who appears named and suddenly four years older at seven-years old in the 1911 Census. Patrick would gain three years of age in only one Census year, recorded as 18-years old in 1911.

Throughout Patrick's life, his vital statistics shift in a fashion I was not accustomed to in this age of personal data. There may have been a tendency to mislead everyone about Patrick's age to help squash the rumors of his origins, but I think lackadaisical adherence to strictly correct ages and spellings was just a sign of those early twentieth century times: Birthdates and ages were not stressed, perhaps even deliberately avoided. For parents on the Reservation, the odds that any one of your kids would make it to adulthood were not all that good—why pin too much on such a feeble hope? Names also often shifted in that era. For instance, the 1909 Indian Census is the only place that I have seen an Indian name for Patrick, transcribed as: *Wa-tam-i-ya-hle*. And this is also the only place that I have seen the last name of "Paul," used for Patrick. Remember, in previous documents his father has a first name of "Pierre" and a middle name of "Paul", but shares a third consistent last name with Patrick. I believe in this one document, Pierre Paul had not finally adopted an Anglicized version of his father's name as his own—and Patrick's—last name. I am not going to use that last name in this book in order to protect potentially living relatives of Patrick from unwanted attention, but that redacted last name, with a couple minor variations (such as a one letter spelling error), is consistently used in all other documents for the rest of Patrick's life.

I was comfortable, in the lengthy research I did for this book, to accept such minor inconsistencies in the details as long as the family ties and structure were a solid match. **I am completely convinced that Patrick Paul *Watamiyahle* is the**

Sasquatch/*Homo sapiens* hybrid that Ed Fusch wrote about.

In the same 1909 census, the 80-year old Catherine or *Klaymeetk*, the widowed mother of Pierre Paul, is shown living in the same household. And the widowed mother of Madeline, Mary Ann *Lakayuse*, also [arbitrarily?] given an age of 80, is also shown in this same, suddenly to us, full household of six.[272] Presumably, both mother and mother-in-law had understood what Hahissiat's pregnancy with Patrick meant and were part of the damage control from 1891 on.

A couple other relatives appear in both the 1909 and 1910 Censuses: Pierre Paul's brother *Jos-is-ken* (or *Jes-is-ken* and sometimes called "George Washington") and his wife, Mary—step uncle and aunt to Patrick. After 1910 I lost the trail of both of them, but they may have left further traces behind for the determined genealogist.

Patrick was now embarking on a remarkably long adult life that would prove to be dramatic, filled with joys, heartache, and strife.

---

[272] June 30, 1909; Colville and Spokane Reservations; Indian Census

Before Patty

CHAPTER FOURTEEN

Patrick Ages 18 to 38

This chapter will trace Patrick's next 18 years, from about age 18 in the last half of 1910, to 38-years old in October of 1930.

According to the 1911 Indian Census, at the declared age of 18, Patrick was now considered a single man, but was still living in Pierre Paul's household. By my timeline, Patrick may have actually been just barely 19 when this census was taken. The records throughout Patrick's life are not going to definitively settle this question of birthdate. I doubt Patrick himself knew when he was born.

I found no mention in the 1911 Census of his grandmothers, Mary Ann (Madeline's mother) and Catherine (Pierre Paul's mother). Mary Ann was still alive in 1911, but apparently not living under the same roof. I found two later references to Catherine.

In the 1912 and 1913 Indian Counts, Patrick is listed separately, living alone, but immediately under the Pierre Paul household. I think he was probably still on the same property, if not under the same roof.

In the 1915 and 1916 Counts, Pierre Paul's household has disappeared altogether (returning in 1917) and Patrick is listed immediately above, but separated from his step-uncle's household. San Pierre (Pierre Paul's brother) was married to

Sophia and by the 1916 census they had four sons and one new daughter (step-cousins to Patrick). The extended clan on Patrick's stepfather's side was apparently numerous, presenting a challenge to any interested researcher to sort out and track down. Patrick was not, however, actually related by blood to this side of the family. I suspect Patrick's mother, Madeline, did have living relatives beyond her aging mother, but I was not able to track down Madeline's tribal affiliation before she married into the Moses-Columbia. Chelan, Nez Percé, Lake, Moses-Columbia, and Yakama all seem like possible tribal connections for Madeline.

I presume that during this time Patrick was honing his farming and ranching skills while working with Pierre Paul.

As you may recall, Dr. Fusch's informants told him that Patrick had no trouble finding a wife because he was a good provider and farmer. A long and eventually troubled marriage began for Patrick on August 24, 1916, when Patrick claimed to be 23 years old (but was really 24), and his new wife, Louise Baptiste (in later years her maiden name is more often given as Jewett), was then 19 years old. Louise is always identified in later documents as full-blood Indian and I have found her identified as 5'-1" and 5'-0" tall in official documents. The Certificate of Marriage from two days before the ceremony was signed by a Catholic priest, a Jesuit, if I understand the initials after his name.[273]

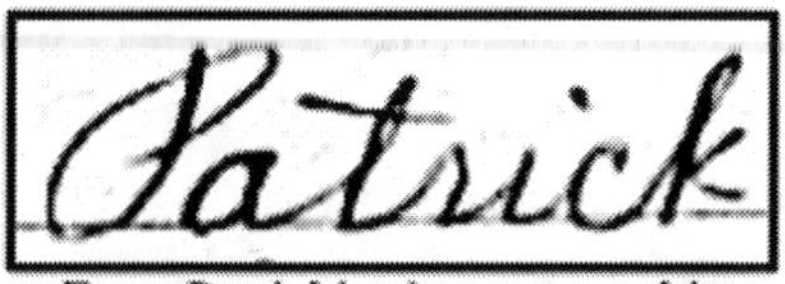

From Patrick's signature on his August 22, 1916 marriage license.

Interestingly, Patrick signed the Certificate of Marriage using his step-grandfather's Indian name as his last name. But the associated Marriage Records have him under the Anglicized version, born in Chelan, as always, with "farmer" as profession. Louise was born in the town of Nelson[274], near Cle Elum, Washington, about two hundred miles away by road.

---

[273] 8/22/1916, Patrick [redacted] & Louise Baptiste; Certificate of Marriage, State of Washington

[274] 8/22/1916, Patrick [redacted] & Louise Baptiste; Marriage Records; Washington State

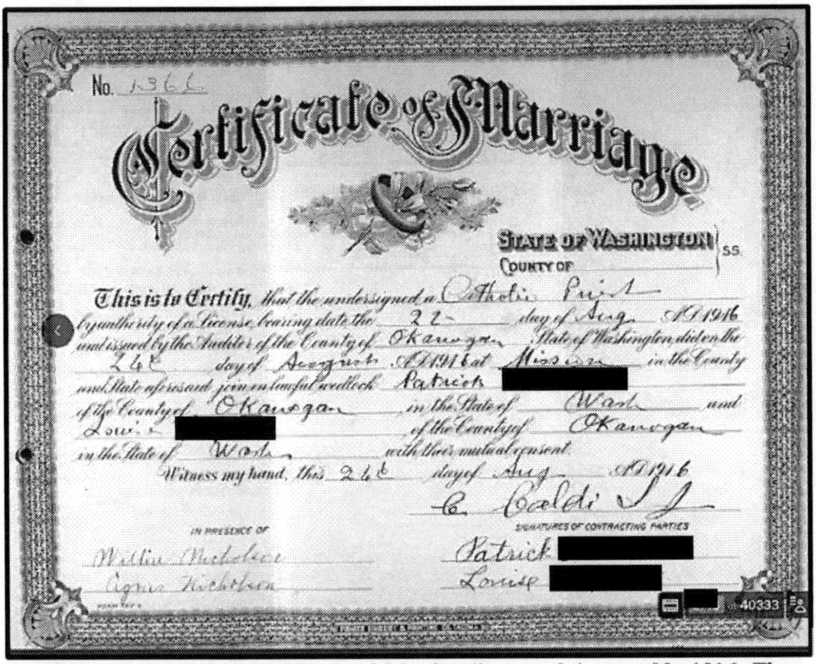

No. 1366

Certificate of Marriage

STATE OF WASHINGTON } SS.
COUNTY OF

This is to Certify, that the undersigned a Catholic Priest by authority of a License bearing date the 22 day of Aug A.D. 1916 and issued by the Auditor of the County of Okanogan State of Washington did on the 24th day of August A.D. 1916 at Miss [redacted] in the County and State aforesaid, join in lawful wedlock Patrick [redacted] of the County of Okanogan in the State of Wash and Louise [redacted] of the County of Okanogan in the State of Wash with their mutual consent.

Witness my hand, this 24th day of Aug A.D. 1916

C. Caldi S.J.

IN PRESENCE OF

Willie Nicholson

Agnes Nicholson

SIGNATURES OF CONTRACTING PARTIES

Patrick [redacted]

Louise [redacted]

[redacted] 40333

Patrick and Louise's "Certificate of Marriage" granted August 22, 1916. The ceremony was performed two days later on August 24th.

Oddly enough, the following 1917 and 1918 Indian Censuses still show Patrick living on his own, with no mention of a wife. My theory is that the people preparing these counts were still often working off of previous counts without interviewing households.

In a document that we will look at later, Patrick is described as a "grain rancher." That is a curious combination of terms. Usually the description would be "grain farmer" or "cattle rancher." I would like to believe that Patrick saw which way the wind was blowing, as did the Colville Indian Agents, and diversified into cattle as a more reliable source of income in the dry climate of the Colville. I believe Patrick was both farmer and cattle rancher. I am confident he kept horses as well. Remember, Ed Fusch reported of Patrick that "he had a good ranch." Ranch is a term usually applied to a place that keeps animals on the hoof.

Patrick joined in when allotments came to the southern half of the Colville less than a year after he was married. When he was 24 years old, in April of 1917, he was granted a Land Patent for

114 acres about seven miles as the raven flies east from the center of Nespelem.[275]

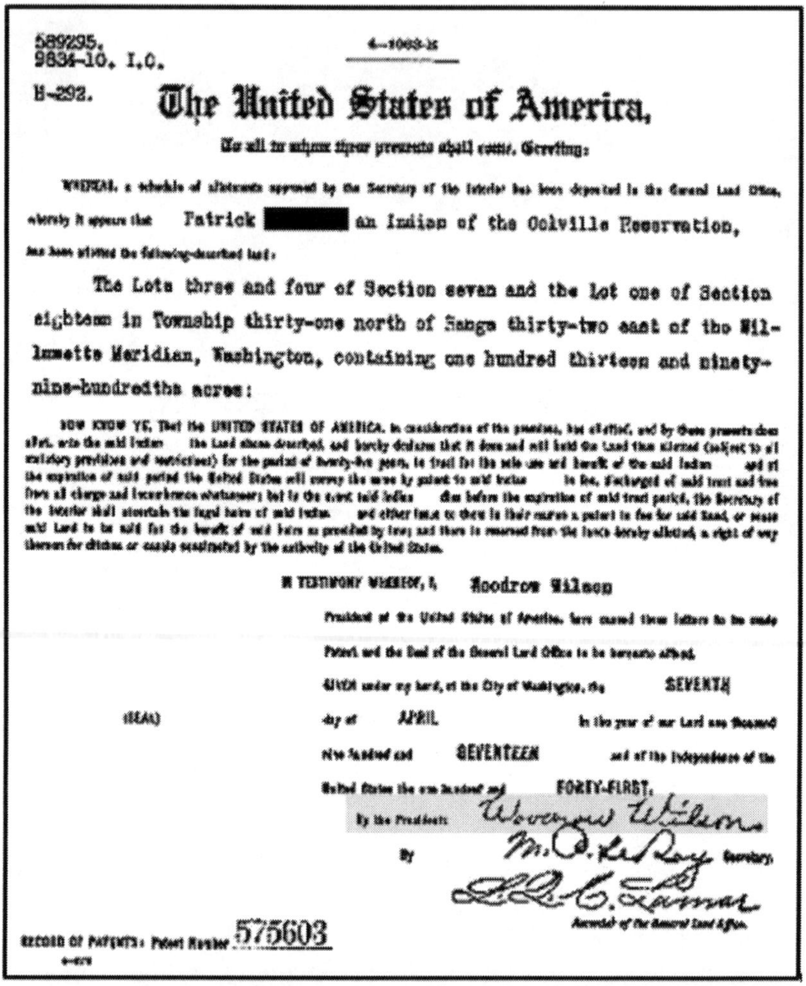

589295.
9834-10. I.C.

H-292. **The United States of America,**

To all to whom these presents shall come, Greeting:

[illegible] a schedule of allotments approved by the Secretary of the Interior has been deposited in the General Land Office, whereby it appears that Patrick [redacted] an Indian of the Colville Reservation, has been allotted the following-described land:

The Lots three and four of Section seven and the lot one of Section eighteen in Township thirty-one north of Range thirty-two east of the Willamette Meridian, Washington, containing one hundred thirteen and ninety-nine-hundredths acres:

NOW KNOW YE, That the UNITED STATES OF AMERICA, [illegible]

IN TESTIMONY WHEREOF, I, Woodrow Wilson

President of the United States of America, have caused these letters to be made Patent, and the Seal of the General Land Office to be hereunto affixed.

GIVEN under my hand, at the City of Washington, the SEVENTH

(SEAL) day of APRIL in the year of our Lord one thousand

Nine hundred and SEVENTEEN and of the Independence of the

United States the one hundred and FORTY-FIRST.

By the President: Woodrow Wilson

By M. P. LeRoy Secretary.

L. Q. C. Lamar

Recorder of the General Land Office.

RECORD OF PATENTS: Patent Number 575603

Patrick's Land Allotment Patent 113.99 acres from April 7, 1917. Note the signature by President Woodrow Wilson. Image formatted for clarity.

That must have been a busy Spring and Summer for Patrick. He had water, some timber, and rich bottom land. It was a start, though not really enough land for a cattle operation in that climate.

---

[275] April 7, 1917, Patrick [redacted]; Colville Indian Trust Patent Number 575603

When I lived in arid southeast Montana—country much like the Colville—I was told by a local rancher that the rule of thumb there was that one cow needed the equivalent of thirty-five acres. It would take 350 acres at that rate just to support ten cows. Patrick may have irrigated some of this land (as the same land is today) and concentrated cows at a higher rate on the rest. More likely though, his cows and steers mingled with a larger communal herd on other allotments and on tribal land. His land devoted to grain, hay, and other crops would have been quickly fenced to keep the horses and cows out.

Some of Patrick's land as viewed from the north. Little Nespelem River in foreground. Google Street View; accessed 2025.10.12

I presume Patrick and Louise continued for some time to live in the shared multi-generational home that the family had been using for years. This family home may have been in the immediate area of this allotment land or closer to Nespelem.

The same month that Patrick got his 114 acres, the United States entered World War I. Only two months later, on June 5, 1917, Patrick had to register for the draft.[276] His Anglicized last name is used, with Nespelem as his place of residence. Birth date: "Sept 1892". He is described as a "Native Am. Indian". Birthplace, as is consistent through dozens of years, is Chelan, Washington. Trade is given as a self-employed farmer.

Listed as dependent upon him for support is, "Wife, father, mother, sister, and grandmother." In 1917, step-father Pierre Paul would have been about in his late forties, Madeline, early forties. His half-sister Christine 13 years old. Whichever grandmother it

[276] June 5, 1917, Patrick [redacted]; Registration Card (Draft Card A) No. 43

was that was still living with them (most likely Mary Ann *Lakayuse*, Madeline's mother), she might have been approaching her mid-eighties. In response to the question on the card: "Do you claim exemption from draft (specify grounds)?" His response was, "Yes, support my family."

On this card Patrick declares no disabilities; he is described as of medium height [a *stretch*], of medium build, with brown eyes, and black hair.

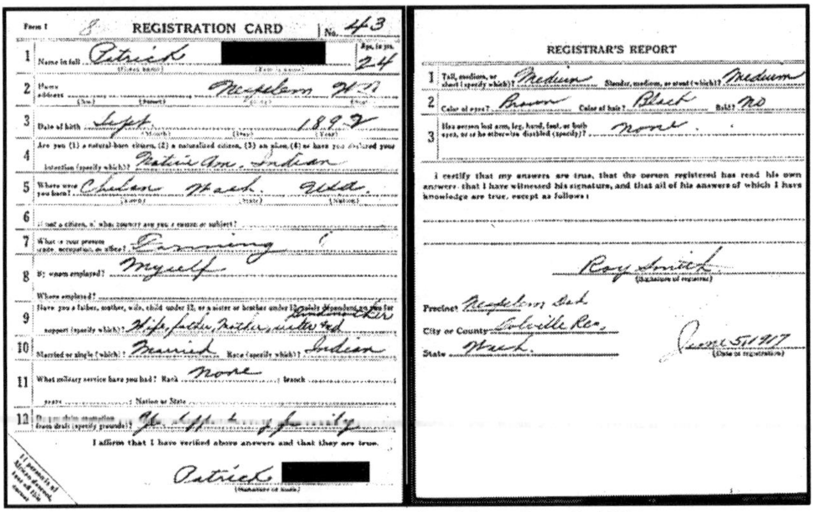

Form 1 8 REGISTRATION CARD No. 43

1 Name in full Patrick [redacted] Age, in yrs. 24

2 Home address Nespelem W.T.

3 Date of birth Sept. 1892

4 Are you (1) a natural-born citizen, (2) a naturalized citizen, (3) an alien, (4) or have you declared your intention (specify which)? Native Am. Indian

5 Where were you born? Chelan Wash. U.S.A.

6 If not a citizen, of what country are you a citizen or subject?

7 What is your present trade, occupation, or office? Farming

8 By whom employed? Myself

Where employed?

9 Have you a father, mother, wife, child under 12, or a sister or brother under 12, solely dependent on you for support (specify which)? Wife, father, mother, sister and Grandmother

10 Married or single (which)? Married Race (specify which)? Indian

11 What military service have you had? Rank None branch

years Nation or State

12 Do you claim exemption from draft (specify grounds)? Yes, support of family

I affirm that I have verified above answers and that they are true.

Patrick [redacted]
(Signature or mark)

REGISTRAR'S REPORT

1 Tall, medium, or short (specify which)? Medium Slender, medium, or stout (which)? Medium

2 Color of eyes? Brown Color of hair? Black Bald? No

3 Has person lost arm, leg, hand, foot, or both eyes, or is he otherwise disabled (specify)? None

I certify that my answers are true, that the person registered has read his own answers, that I have witnessed his signature, and that all of his answers of which I have knowledge are true, except as follows:

Roy Smith
(Signature of registrar)

Precinct Nespelem Sub.

City or County Colville Res.

State Wash.

June 5, 1917
(Date of registration)

Patrick's World War I Draft Registration Card from June 5, 1917.

The signature on the card looks different from what is on his marriage license, although the signatures were generated only a couple month apart. I trust that the marriage license signature is more likely to be Patrick's own. On that form, the signature of Louise and the two witnesses all differ from each other. On the draft card his signature looks much like all of the handwritten answers throughout the form. Perhaps Roy Smith, the registrar, filled out the form on Patrick's behalf.

Finally, in the 1920 U.S. Federal Census, [Pierre] Paul is still listed as the head of the household, with Patrick as a married dependent, sharing the house with his mother, Madeline, and half-sister, Christine. There are two family member changes since the 1917 draft card: No grandmother is included, but there is also the happy addition of a one-year old son for Patrick and Louise, apparently their first child, identified then only as "Baby [redacted] [Patrick's last name]".

On the 1922 Indian Census, there is a note that the full-blooded, Mary Ann *Lakayuse*, born in 1829 (take that date as approximate), had died in 1920.[277] This was Madeline's mother and perhaps the last living link to the days before the white man changed everything.

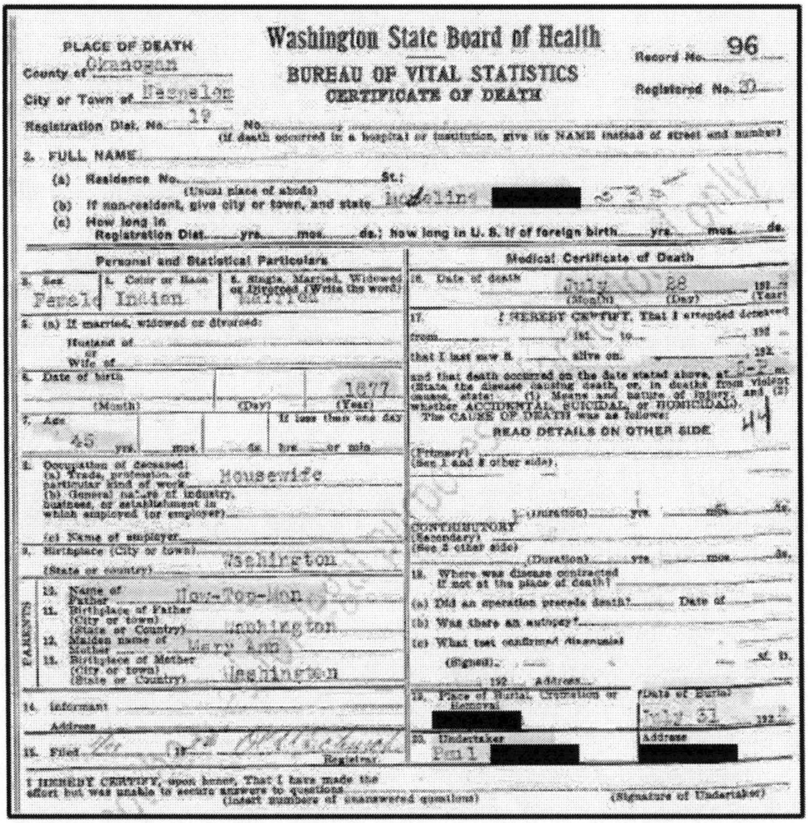

Washington State Board of Health
BUREAU OF VITAL STATISTICS
CERTIFICATE OF DEATH

PLACE OF DEATH
County of Okanogan
City or Town of Nespelem
Registration Dist. No. 19
No. (If death occurred in a hospital or institution, give its NAME instead of street and number)
Record No. 96
Registered No. 30

2. FULL NAME
(a) Residence No. St.; (Usual place of abode)
(b) If non-resident, give city or town, and state Madeline [redacted]
(c) How long in Registration Dist. yrs. mos. ds.; How long in U. S. if of foreign birth yrs. mos. ds.

Personal and Statistical Particulars

3. Sex: Female
4. Color or Race: Indian
5. Single, Married, Widowed or Divorced (Write the word): Married
5. (a) If married, widowed or divorced: Husband of or Wife of
6. Date of birth (Month) (Day) (Year) 1877
7. Age 45 yrs. mos. ds. If less than one day hrs. or min.
8. Occupation of deceased: (a) Trade, profession, or particular kind of work Housewife
(b) General nature of industry, business, or establishment in which employed (or employer)
(c) Name of employer
9. Birthplace (City or town) (State or country) Washington

PARENTS
10. Name of Father How-Top-Mon
11. Birthplace of Father (City or town) (State or Country) Washington
12. Maiden name of Mother Mary Ann
13. Birthplace of Mother (City or town) (State or Country) Washington

14. Informant
Address
15. Filed

Registrar

Medical Certificate of Death

16. Date of death July 28 (Month) (Day) 192_ (Year)
17. I HEREBY CERTIFY, That I attended deceased from 192 to 192 that I last saw h alive on 192 and that death occurred on the date stated above, at m. (State the disease causing death, or, in deaths from violent causes, state: (1) Means and nature of injury, and (2) whether ACCIDENTAL, SUICIDAL, or HOMICIDAL). The CAUSE OF DEATH was as follows:
READ DETAILS ON OTHER SIDE
(Primary) (See 1 and 2 other side).
(duration) yrs. mos. ds.
CONTRIBUTORY (Secondary) (See 3 other side)
(Duration) yrs. mos. ds.
18. Where was disease contracted if not at the place of death?
(a) Did an operation precede death? Date of
(b) Was there an autopsy?
(c) What test confirmed diagnosis?
(Signed) M. D.
192 Address
19. Place of Burial, Cremation or Removal [redacted]
Date of Burial July 31 192_
20. Undertaker Paul [redacted]
Address [redacted]

I HEREBY CERTIFY, upon honor, That I have made the effort but was unable to secure answers to questions (Insert numbers of unanswered questions) (Signature of Undertaker)

Death Certificate for the woman that bore a Skanicum/H. sapiens hybrid, Patrick's mother: Madeline Hahissiat Kolockun. Born ~ 1875 and died July 28, 1922

This began a period in Patrick's life of great changes, deaths, and births. Perhaps the harshest blow of all was the death of his mother, Madeline on July 28, 1922 at 5:00 PM at the estimated

[277] Indian Census; Colville and Spokane Reservations; June 30, 1922

age of 45 (she was about 16 when she was abducted by the Skanicum). Patrick was 29 years old in my estimates. I have not been able to find a cause of death for Madeline *Hahissiat*, but the info I do find confirms her as kin to Pierre Paul, Patrick, and Christine. And even provides "Mary Ann" as her mother's name. The Death Certificate lists [Pierre] Paul as the undertaker. She was buried on July 31, 1922.[278]

Madeline protected and raised the hybrid child Patrick to adulthood and I think he was still living in the same house or cluster of houses with her when she died. The effect of her death on his sense of well-being must have been enormous, particularly with the new responsibilities that he was then taking on.

Another family head with the same last name as Pierre Paul, the approximately 63-year-old Herman [redacted], died on April 22, 1923, leaving his 47-year-old widow, Elizabeth, and two young daughters, Alice M. and Madaline.[279] I presume that this family was some sort of kin to Pierre Paul. More importantly, they were soon to become even closer relatives to Patrick.

More change came with the birth of Patrick and Louise's second child on or about September 1, 1923—Patrick's second son, Ernst. It is a few years and a couple chapters before we will hear more about Ernst, but his life also was a dramatic one, and he was at times very close to his father.

In 1924, Pierre Paul, perhaps in his sixties (year of birth varies widely for Pierre Paul), married for the second time to the Elizabeth who was widowed when her husband died in 1923. Elizabeth was somewhere in her forties and brought along her two children: Alice M., who was born in 1912 (about twelve years old when she joined Pierre Paul's household), and the ironically named "Madaline" and sometimes "Madeline" who was born in 1915 (about nine years old). I could find no marriage license or certificate for Pierre Paul and Elizabeth. This marriage may have been more one of support to a distantly-related widow with children than a formal legal union.

---

[278] 7/31/1922, Madeline [redacted]; Certificate of Death; Bureau of Vital Statistics; Washington State Board of Health

[279] Indian Census; Colville and Spokane Reservations; June 30, 1923

By June 30, 1924, Pierre Paul's brother, San Pierre, seems to have died[280], leaving another widow, this time with four children—the oldest at fourteen years old and down to toddler Basil at only one year old. Presumably, Pierre Paul and Patrick also had much of the responsibility for keeping this mourning family fed, clothed, and housed.

Pierre Paul's second "marriage" to Elizabeth did not last long. Elizabeth died the very same year on December 31st, 1924, but Alice M and young Madeline continued to live in Pierre Paul's house.[281]

Tragically, at the start of 1925, Christine, at 21 years old, was diagnosed with pulmonary tuberculosis. She was likely still living in the same house as Patrick, his wife Louise, and their approximately six-year-old son, graduated from "Baby [redacted]" and now identified by the first and middle names, Simon Brooke.[282] Unfortunately, before the year was over, Simon Brooke would die of unknown causes at six years old. Living in the same house as someone with tuberculosis provides the sustained close contact required for TB transmission. It seems likely that he died of tuberculosis or perhaps succumbing to pneumonia as a secondary infection to TB. It's even possible that Simon Brooke contracted TB first and gave it to his caregiving Aunt, Christine.

It is pure speculation how Patrick took the death of his son and first child, Simon Brooke, when Patrick was about 32 years old. Perhaps this is when a persistent problem with alcohol took hold and would soon dominate the last decades of his life.

At about the same time as Simon Brooke's death, came the birth of another child to Patrick and Louise on September 25th, 1925: Marie, sometimes called or misspelled Masie. Marie was their third child and first daughter.

Obviously, the 1892 hybrid, Patrick, was fertile. These children answer the question of whether Patrick's birth was just a freak, brief, and highly unlikely introgression of Sasquatch DNA into *Homo sapiens* with a firm, "No!" An introgression is when

---

[280] Indian Census; Colville and Spokane Reservations; June 30, 1924
[281] Indian Census; Colville and Spokane Reservations; June 30, 1926
[282] Indian Census; Colville and Spokane Reservations; 1925; Deaths

genes from one organism are introduced into another species, in this case through hybridization. But these two boys and one girl were apparently healthy *and* had 1/4 Skanicum blood and genetics, born about the time that my parents were born. Children of theirs, 1/8 Skanicum blood, would be about my age now. I am happy to report that I am still alive and well as of this writing. My living example implies that Patrick may have grandchildren, with *1/8 Sasquatch DNA*, alive and well as of this writing, enjoying occasional lattes, doting on their grandchildren, and grappling with recent retirement.

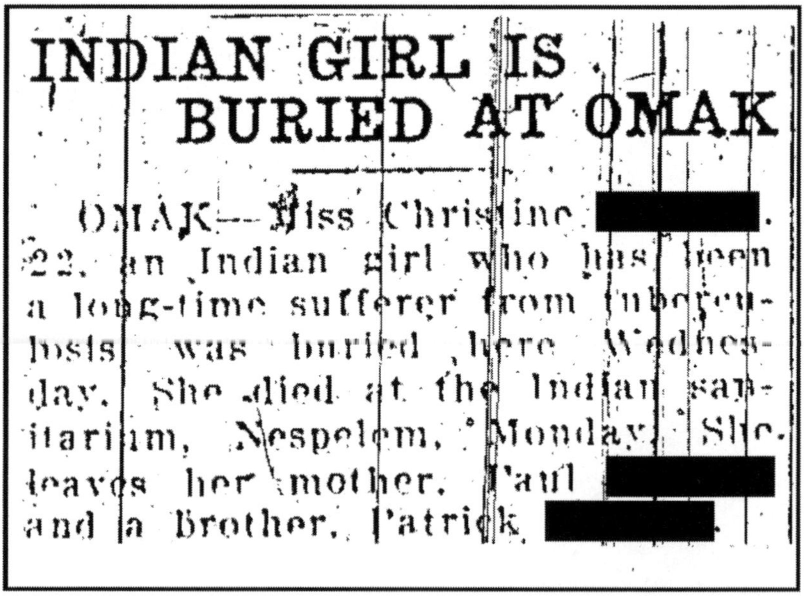

INDIAN GIRL IS BURIED AT OMAK

OMAK—Miss Christine [redacted], 22, an Indian girl who has been a long-time sufferer from tuberculosis was buried here Wednesday. She died at the Indian sanitarium, Nespelem, Monday. She leaves her mother, Paul [redacted] and a brother, Patrick [redacted].

Obituary for Patrick's half-sister, Christine, published in early July, 1926 in the *Wenatchee Daily World*. It is interesting that her father, [Pierre] Paul, is not specifically identified as such.

It was a second devastating blow for Patrick when his half-sister, Christine, died June 28, 1926 of the tuberculosis that she had lived with for 19 months.[283] Patrick had known Christine her entire life and, outside of his mother and his children, Christine was the only blood relative that he knew. The Certificate of Death

[283] June 28, 1926, Christine [redacted]; Certificate of Death; Bureau of Vital Statistics; Washington State Board of Health

shows her as single, 22 years old, born in 1904, with no doctor in attendance at her death. The document does list both father, Pierre Paul, and mother, Madeline, as born in Washington state. That is filling in another missing piece of information. Some of the Indians on the Colville had ranged up into Canada, over into Idaho and Montana, and down into Oregon. Undertaking services were provided by "friends"—the norm in those days on the Reservation—washing and dressing the body for burial.

Less than a year after his half-sister, Christine had died, Patrick and Louise had their fourth child, Stella Lucy, on June 7th, 1927.[284]

We hear no more of Patrick for over a year, until October 1928. I am concerned that Patrick was developing a growing personal problem after the deaths of his mother, Madeline, in 1922; his first child, Simon Brooke, in 1925; and the death of his half-sister, Christine, in 1926.

One day, when I thought I had surely exhausted every avenue of dredging-up information regarding Patrick's life, I decided to give it one more shot (there has been about a dozen "one more" shots since). I searched for Patrick using his grandfather's name as a substitution for Patrick's Anglicized last name. And I found something—an arrest record from October 10th, 1928. An unfortunate incident that was also a gift for all those who want answers to the ancient mystery of Sasquatch.

[284] Indian Census; Colville and Spokane Reservations; June 30, 1928

CHAPTER FIFTEEN

What is Sasquatch?

I enjoy the podcast, *Sasquatch Chronicles*, with Wes Guermer. I really like it very, very much. I must have heard a thousand episodes. Wes does a great job of letting people talk and I have grown in my ability to flexibly consider the relict hominid subject as Wes has grown in his understanding. As he implies: The old ways of trying to solve the mystery have not gotten us very far after fifty-plus years.

One thing Wes does as he is wrapping-up with a guest, is ask, "What do you think Sasquatch is?" I have rehearsed my simple answer to the question just in case he ever has me on the show to recount my non-dramatic experiences. "Sasquatch is the sum total of all that your witnesses have observed." This chapter attempts to answer Wes' question by looking at what witnesses are telling us that Sasquatch is. The answer is not as simple as, "Sasquatch is an elusive, bipedal mountain-ape." Or "Sasquatch is a shape-shifting, interdimensional alien." Or "Sasquatch is a relict hominid; a ninja of the forest at keeping hidden."

Sasquatch, like the humans we normally hang out with, are very complicated and their complex natures not easily reduced to a phrase or a sentence or a paragraph. The question deserves at least a chapter as we try to understand what Patrick himself inherited through his patrilineage.

> They always seem to do everything for a reason, wasted no time on anything they did not need. When they were not looking for food, the old man and the old lady were resting, but the boy and the girl were always climbing something or some other exercise.[285]

"They always seem to do everything for a reason" is what Albert Ostman concluded after his 1924 abduction. Witnesses often report Sasquatch as seeming to be on a mission as they continue walking in a straight line, sometimes completely ignoring the shocked Hairless Ones nearby.

Ideas about their growth rates and lifespans are all over the place. Alley suggested that females did not develop secondary sex characteristics (presumably breasts and wide hips) until they are about six feet in height.[286] Fahrenbach is one researcher that had enough data to make some conclusions, particularly about the maturation of young females:

> …I have received three sets of footprint measurements (from three different observers), each stretching over several years, that were considered to originate from three growing animals… The graph suggests a rapid growth from the sasquatch such that a juvenile animal, age 8 years, with a foot length of about 9" (23 cm), would be about 5'9" to 6' in height (173-183 cm), in effect the size of an adult human male. …offspring seem to be spaced at intervals of about 5 years… records of family groupings of prints suggest that an older sibling will stay with the mother for some years after the arrival of the next one, until about age 10… The female probably reaches sexual maturity at about age 9-10… by which time she will be near 6' (1.83 m) in height. At this stage, the female has been observed to keep male company, and has developed small breasts.[287]

---

[285] Bigfoot Research Organization (BFRO) Report #1091 2000; https://www.bfro.net/GDB/show_report.asp?id=1091
[286] Alley 2003 (2018 edition), page 299
[287] Fahrenbach 1997-1998

Such a rapid growth rate further confirms my suspicion that most of the frightening road crossings, chance encounters, and campsite harassment are really just brushes with teenagers, still learning how to negotiate their world and how to steer clear of the Hairless Ones. Seemingly horrifying run-ins from our perspective may just be thoughtless teasing and a test of their growing abilities.

The number of individuals in an extended family group or clan varies widely in the few accounts that claim to be able to assign a number to such things. There are Sasquatch reported that appear to be living entirely on their own (generally old or "rogue" males). In the Ostman account, it appears that four Sasquatch in British Columbia (in 1924) were genetically isolated. Joan Ocean specifically reports 33 for a "family" in the Kiamichi Mountains of Oklahoma.[288] On the high end, I know a man in upstate New York who says his local clan has over 70 individuals. Group counts, and the relationship ties behind them, are very difficult to establish with certainty. Generally, unique tracks, vocalizations, and sightings support a number range of one to about ten individuals within a specific area. The limited observations we have do not negate the possibility that these smaller family groups are socially, linguistically, culturally, and genetically grouped to larger populations in the surrounding regions.

The little *anecdotal* information that we have on mating is that of young adult Sasquatch of either gender leaving their home territory to find mates. Bland reports of a case in which a female left British Columbia, but returned with a male mate from Oklahoma.[289] In another case, Noël recounts a story received *telepathically* by a high-functioning autistic teenager in Iowa. As that story goes, two young males set off looking for a mate for one of them. They traveled up to the "north country" before returning with a female.[290]

Sasquatch are rarely seriously aggressive, though just a throat-

[288] Morehead 2017 (second edition), page 217
[289] Bland 2022, p 319
[290] Noël 2019, page 264

clearing with their deep voices and powerful lung capacity may come across as horribly frightening and angry. Alley notes:

> It is somewhat remarkable that, in spite of periodic doses of flying lead, sasquatches do not seem to be very good at measuring up to their ferocious tabloid image... It appears that sasquatches are as difficult to anger as the gentlest of the great apes, the gorilla...[291]

That's not to say they are not capable of asserting their needs or that they have no red lines we shouldn't cross. In Alaska, there are rumors of no-go areas where miners and hunters disappear with enough frequency that these places gain a reputation as too risky to venture into.

I know of two small communities in Alaska (one on the Yukon river, one coastal) supposedly abandoned under pressure from what up here is called the "Boreal Bigfoot" (coined by Michael Thompson of Sasquatch Tracker).

Pyle writes from Washington state: "On the western flanks of Mount Adams… Council Lake, at the bluff's base was long ago abandoned, some Klickitats say, because of numerous Bigfeet near there."[292] Pyle continues later in the book with a passage describing how Jorg Totsgi, editor of *The Real American* out of Hoquiam Washington in 1924, wrote that Sasquatch/Seeahtlk was harmless unless antagonized, but vengeful if abused—killing twelve Indians for every Seeahtlk who was killed.[293]

Isdahl received a shocking letter with a story about an ancestor in the early 1900s violating a no-go zone in the state of Virginia in the U.S. The man was found dead, having committed suicide by his own rifle, but not before both his legs had been broken. This was an area that the family considers a no-go zone to this day because of the belief that Sasquatch might kill you if they find you in there.[294]

Lapseritis agrees that Sasquatch are "fiercely protective" of the

---

[291] Alley 2003 (2018 edition), page 48
[292] Pyle 1995, 2017 edition, page 101
[293] Ibid., pages 154-155
[294] Isdahl 2021, page 78

private areas where they live and have families.[295] Some of that impression of fierceness is confirmed through one of Noël's witnesses—Denise in Maine—when she describes strong mental impressions that she was getting from a female Sasquatch:

> And then I was feeling *his* [the alpha male of a Sasquatch family] energy through *her* [the alpha male's mate]... It was pictures where the protector [the alpha male] sees men coming in with guns, and he will protect his family and do what he must do. …she didn't want me to see what he would do.[296]

Often aggression from Sasquatch seems to have no noticeable trigger or cause. Perhaps in this next case—that Isdahl reports from northwest Georgia—the trigger was the noise of an all-terrain vehicle (ATV) simply interrupting a Sasquatch hunt or even a nap. The moving ATV was flipped from rear to front with two riders on it. As they were scrambling to check for injuries and roll the ATV back up again, they heard a roar. The men hopped back on the ATV and tried to get out of the area as quickly as possible, but found themselves pursued by something very large on all fours. One of the men fired a few shots from a pistol before the quadruped broke off pursuit.[297]

Guermer covered a secondhand oral account, originating in April 1951, from one of the investigating police officers. In this case, a farmer was thrown against a metal gate by "a big blond booger" (as the farmer described it). The farmer spent at least the next 6-8 weeks in a hospital and it was uncertain if he ever recovered from his injuries. Booger motivation unknown.[298]

Someone shared a 1933 diary entry with Short of an incident in which a "wild man" tore the hide from a pack mule, killed a dog, and bruised a miner during a snowstorm on the east slopes of Mt. Whitney.[299] Motivation also unknown.

A young man was on his annual clandestine trip to gather wild

---

[295] Lapseritis 2011, page 74
[296] Noël 2019, page 39
[297] Isdahl 2020 pages 176-178
[298] Guermer 2025, show # 1178
[299] Short 2024 pages 109

marijuana near Warsaw, Indiana. While hiding from humans in a corn field, he hears intimidating noises. Running to escape the maker of those noises, he was plowed over from behind by something *big*. He did not get a look at his pursuer. He was pushed into the mud so deeply that he had trouble getting up when he finally regained consciousness at least eight-hours later. As reported to Isdahl.[300]

Short again: One night, a guard at the Twenty-Nine Palms Marine Base in California made the mistake of pointing a rifle at a Sasquatch. The Sasquatch rushed the guard, grabbed the rifle, and bent the barrel.[301] Amazingly, the guard received no serious injuries. Moral of that story: Do not point your rifle at a Sasquatch.

A woman hunting in the swamps of south Louisiana heard a “weird low humming sound" and then saw two female Sasquatch, one with a baby on her back. The witness took a photo of the three with her iPad and then headed out of the swamp for home. That night a big male Sasquatch “raised hell” at her house and continued to do so nightly for weeks, as transcribed in Isdahl’s book.[302] One of the “rules” Sasquatch seems to operate under is, “Thou shalt not allow thyself to be photographed.” Seriously. That seems to be a rule.

Alley uncovered two reports from Southeast Alaska of thrown snowballs (up to 10” in diameter!) and other reports of occupied vehicles shaken.[303]

In northern Canada, Grossinger reported that a man stopped at a highway pullout and noticed a dead animal smell. He ventured into the woods to find the source of the smell and a Nahganne (a Yukon Territory term for the Boreal Bigfoot) eating off of an animal carcass. The Sasquatch began walking towards the man, “arms swinging” threateningly. The man ran back to his vehicle and drove off.[304]

Aggression, seemingly frequent by reading these previous accounts, is actually quite rare, but sometimes the rare aggression

---

[300] Isdahl 2021, page 162
[301] Short 2024 page 219
[302] Isdahl 2021, page 145
[303] Alley 2003 (2018 edition), pages 216-220
[304] Grossinger 2022, page 135

is just plain *gross* as witnesses were to discover in the southern Washington State Cascade Mountains when an unseen Sasquatch urinated on the front and back doors of their tent, as published by Isdahl.[305]

Isdahl points out, humans may be so terrified when confronted by an aggressive Sasquatch that they will lose control of their own bladders and/or bowels.[306]

The recently passed dedicated Yukon researcher, Red Grossinger, personally experienced some aggression in July 2011: "...large trees being thrashed about…with low guttural growls, seemingly in an attempt to chase me away. (And it worked…)"[307]

In the 1970s in the Vyatka region of Russia, forester Boris Liberov, encountered two females with a young one in a berry patch 50 meters away. The hairy beings threw a root at him to drive him off. That worked! He left![308]

Baxter reports that on the Seward Peninsula of Alaska "…the Nantiinaq was said to be both elusive and deadly."[309]

Yet, experiences may be anything but aggressive. As I mentioned, all of my experiences have been very gentle, to the point of subtle. Noël quotes Sophie of Nova Scotia recalling a childhood encounter:

> You hear these people that try to paint Bigfoot as something that would have killed me, eaten me or kidnapped me, but I didn't get any of those vibes from them whatsoever. They were like humans. If anything, I got a strong sense of intelligence, warmth, kind hearts…[310]

And Noël again: "They are certainly capable of being aggressive and terrifying... but on the whole, they are far more monotonous than monstrous."[311]

---

[305] Isdahl 2021, page 150
[306] Ibid., page 81
[307] Grossinger 2022, page 194
[308] Burtsev 2015 page 151
[309] Baxter 2021
[310] Noël 2019, page 55
[311] Noël 2023, page 168

However, Skanicum does give the overall impression of having anger management issues that involve lots of screaming. Wiitala gave a personal example of what seems to be an overreaction: "Later that night I had to get out of my tent to relieve myself and what sounded like a ten-year-old boy screamed at me from the woods."

Lapseritis explains that they will threaten to kill when they feel threatened themselves, but will soon go back to a "normal" mood and don't hold a grudge over perceived slights.[312]

Modern human babies and children generally get a pass on this emotional volatility. According to Lapseritis, Sasquatch will seek out children just to watch them.[313] Parents of children naturally find this behavior of watching their kids from concealment in the woods unsettling. Paulides finds that Bigfoot are not scared of children (though they often act frightened of modern human adults) and cites one case of a 15-year-old autistic boy aware of being watched as he worked outside. The boy was scared by this behavior but somehow knew that he would not be hurt.[314]

Grossinger wrote of a 7' tall being covered in dark hair making repeated visits to a girl's bedroom window. It would look at her, smile, and wave in a friendly manner.[315]

There's nothing creepy about that!

Laughter, from any age of human, seems to be another attractant. If you want to stir up some Sasquatch action, find a place with good wood cover and have a bonfire at night with cheerful friends. You might soon have that sensation of being watched.[316]

Although I cannot recall a case of anyone seeing a Sasquatch defecating, urinating, making stick structures, or having sex—though they presumably do all of those things—there have been lots of visuals of them drinking and eating.

Marlowe recounted the story from 1955 of William Roe on Mica Mountain near Valemont, British Columbia, watching as a

---

312 Lapseritis 1998, page 45

313 Ibid.

314 Paulides 2017, pages 387 & 407

315 Grossinger 2022, page 211

316 Lapseritis 1998, page 45

female Sasquatch, covered all over with no more than 1" of hair (no hair around her eyes), using her white and even teeth to pull and eat leaves off of a bush. Roe was close enough to see that her eyes were small and dark and that she had human-looking ears. After the Sasquatch sensed him and had left, Roe had the presence of mind to search for scat. He found some that he believed belonged to the mysterious creature he had just seen ("mysterious" because in 1955, Bigfoot had not entered popular awareness). The scat contained no hair or insect shells and Roe concluded that the animal was eating a plant-based diet.[317]

Paulides shared a story from northern California very similar to the Roe report. This time the Bigfoot was pulling willow branches through its teeth to eat the leaves.[318] And a Sasquatch harvesting cattails from a swamp was reported to Isdahl.[319]

Burtsev included pictures and descriptions of his find of gnawed aspen branches and trunks at 150-170 cm (60-67") above the ground at the scene of filmed tracks and a Sasquatch sighting in the Kemerovo region of Russia in 2013.[320] Boise Bigfoot produced a fine short clip that includes this Russian video (see the link in the footnote).[321] Igor "Two Hoondred Dollah" Burtsev is second from the left in the scenes of the boys watching the video while seated at a table. Note also the very large man on the extreme left, next to Burtsev. That is Nikolai Valuev, the tallest (at 7'-0") and heaviest (peaking at 333 pounds) World Boxing Association world champion—twice! He had a career record of 50 wins in 53 bouts—before stints in television and movies—and he is now a Russian politician. He became interested in relict hominids after a boxing spectator jeered at Valuev that Valuev was himself a Yeti. He attended the 2011 International Forum of Hominology in his home country of Russia along with Burtsev and such notables as the late Dr. Jeff Meldrum, Dmitri Bayanov, Ron Morehead, and Dr. John Bindernagel (also unfortunately

---

[317] Marlowe 2013 pages 25-26

[318] Paulides 2017, page 418

[319] Isdahl 2020, page 71

[320] Burtsev 2015 page 391-392

[321] Boise Bigfoot, https://www.youtube.com/watch?v=sfrjLh8jNWo; accessed 2025.10.15

deceased). Please note in the video at 0:17 the unusual profile of Valuev's skull, particularly that of his brow ridges, the extreme slope of his forehead, how his skull peaks to the back, and the misleading effect of a proportionally small braincase. In actuality, Valuev's body is so large that he just appears to be small-brained by comparison. In group photos in the Burtsev book it is easy to see that Valuev's brain is every bit as big as those of the other puny humans around him. The gray matter is just lost behind a very large face.

Valuev has acromegaly, a condition of excess growth hormone. The jeering spectator had a point: There are similarities between Valuev's cranial profile and what is typically reported for Sasquatch. And his skull has similarities to Patrick's that we will explore in the next chapter. Valuev might not be a Yeti, but his existence is a hint that there is little, genetically, separating Skanicum from human. There may be other elements behind the differences in morphology between Sasquatch and "modern human" besides DNA.

Fahrenbach makes an interesting point about Sasquatch nutritional needs when he writes:

> ...[According to] Kiciber's Law (MacMahon and Bonner 1973), which states that the basal metabolic rate scales as the ¾ power of mass, a massive animal needs less energy input per gram of body weight than a small one does. This means that a Sasquatch can get by with a *relatively* smaller amount of food than a smaller animal.

Fahrenbach calculated that a 650 pound (300 kg) Hairy One would need about 5,000 calories per day, rising sharply in bad weather or with exercise. He postulated that in winter, it might be necessary for Sasquatch to shift to a more carnivorous diet to supply all the calories needed in spite of the discount from Kiciber's Law.[322]

Continuing our brief survey of likely Skanicum eating habits, Paul in West Virginia, according to Noël, observed a girl and even

---

[322] Fahrenbach 1997-1978

younger boy Sasquatch eating mushrooms, grubs, and young catfish. One catfish was dramatically caught by hand when the girl Sasquatch plunged her fist through the clear ice of a frozen lake.[323]

Isdahl, quoting Short on his *How to Hunt* podcast of 11/03/2025, read separate accounts of witnesses observing or seeing the evidence of Sasquatch eating Saskatoon berries, alder leaves and bark, and spring jack pine needle-tips. Alley noted water lily tubers, skunk cabbage, and even devil's club stems and roots.[324]

Burtsev reports on food taken by Sasquatch from "feeding stations" as including oats and other cereals, honey, chocolate, and candy. He expands:

> It turned out, that snowmen really like chocolate… As for meat, snowmen do not eat it very actively. It was noted that after killing such animals as goats and sheep, they first ate the liver, heart, and kidneys.[325]

However, Sasquatch in North America seem happy to eat both meat and animal organs, and have a particular fondness for pig—wild or domestic. Isdahl published one story of a 150-pound pig being cleanly snatched out of a pen with 4-5' walls.[326] The same author included a story of a witness seeing a 6' Skunk Ape in the Everglades eating an alligator leg. Oddly, this Skunk Ape appeared to have a necklace or *tracking collar* on it.

Over much of the North American Sasquatch habitat, various species of deer are likely to be a staple, when they can be caught. Fahrenbach helps to clarify the mystery of deer (*Odocoileus* species) found with their torsos opened up and internal organs missing. The internal organs are themselves high in protein and nutrients. Choice organs would make a tender raw meal for an animal with dentition similar to ours. The rest of the carcass may be deliberately left behind for a few days to, "…ripen the meat

---

323 Noël 2019, pages 100-101
324 Alley 2003 (2018 edition), page 295
325 Burtsev 2015 page 232
326 Isdahl 2020, page 50

and soften its texture…"[327]

Alligator is not the most peculiar thing Sasquatch has been seen eating. There are plenty of encounters reporting dumpster diving, even behind suburban apartment buildings at night. Still less appealing is one of the classic stories passed around in the Sasquatch researcher world and included by Paulides. The story is that of a near 10' tall Bigfoot caught on videotape in about 2001 at Oklahoma's Cheyenne and Arapaho Lucky Star Casino. The big guy was raiding a commercial kitchen grease pit.[328] With personal experience of kitchen grease traps, I can affirm that the contents are *extremely* nasty.

Bigfoot generally have a poor or even predatory relationship with dogs. Their general dislike of dogs may be simply because dogs have extremely keen hearing and smell—they know when other animals are around long before us weak-sensed humans have a clue. I often use a dog on a long leash in front of me as my early warning system as I walk through woods in grizzly, black bear, and moose country. I just watch the dog for a turning head and perked-up ears. It saves my own head from a lot of swiveling around.

Although the usual reaction by dogs to just *smelling* Skanicum is to retreat, whimper, hide, refuse to move forward, run back to the truck, etc., aggressive dogs sometimes have the upper hand over young or timid Sasquatch. Isdahl repeated a case of a witness seeing five or six dogs chasing a slender-looking seven to eight-foot-tall Sasquatch through a sleeping neighborhood on the outskirts of Bellingham, Washington in about 1974.[329]

Sometimes individual Sasquatch develop a habit of preying on dogs for food. One odd piece of evidence I personally stumbled across in the Alaskan boreal forest half a mile from the nearest road or trail was that of *half* of a domestic dog. This was the lower half, starting just below the ribcage, neatly separated from the front of the unfortunate dog. It had been a smallish dog, maybe only 20 pounds when one-piece. My efforts to find the dog's

---

[327] Fahrenbach 1997-1998

[328] Paulides 2017, pages 401-402

[329] Isdahl 2021, page 96

owner failed. No local seemed to be missing a dog. There is a busy highway about a mile from the grisly location. The dog may have come from a camper or was callously abandoned on the roadside. To this day, I call that area, "Half Dog Ridge." I make a point of keeping dogs close or on leash in the woods around here.

Redfern reports in his 2016 work, *The Bigfoot Book: The Encyclopedia of Sasquatch, Yeti, and Cryptid Primates*, that the bipedal Yeren of Central China hunt and kill dogs routinely.[330] Paulides, among many, reports a normally fierce dog cowering and shaking in the presence of a Sasquatch.[331]

In December 2023, in personal correspondence to a Canadian Sasquatch enthusiast, I wrote on the subject of Sasquatch and dogs:

> Sasquatch has a mixed relationship with predators suggesting that Sasquatch has an enormous range of cultural tendencies just like you would expect to find in indigenous peoples all over the world. You can certainly find Australian Aborigines and the Inuit valuing canines for companionship, guard animals, hunt helpers, and working dogs, but you can also find humans that kill canines on sight as a pest and even see them as a food source. You may find people in partially wooded rural areas having a hard time hanging onto dogs—they just keep disappearing. In this spacious subdivision I live in now, we have the occasional dog turn up gone. I have heard a regional story of a dog musher coming home to find no more dog team. These dogs were on chains in a dog yard and they were freight-hauling dogs, which means that they were big dogs, pushing 100 U.S. pounds (45 kg). Their disappearance was chalked-up to wolves, but I very much doubt that wolves were involved. Even a large wolf pack is unlikely to haul off an entire collared dog-team at one go. Then just last night I listened to a *Sasquatch Chronicles* story in which a hunter was run off of his deer kill by a Sasquatch. He returned two days later to find a hunting dog petrified in fear—too scared to

[330] Redfern 2016, pages 334-337

[331] Paulides 2017, pages 155

move—but so happy to see him that it peed itself. The deer was gone, but the dog was in its place. Very strange. Stranger yet was that the tag on the collar helped to locate the dog's owner TWO COUNTIES AWAY. But this dog was not beat up and scrawny from a long, hungry journey. The witness implied that he thought the dog had been carried to where he found it. As a pet? As food? Sasquatch has even been reported to be seen *petting* dogs. Dog collars have also been found amongst gruesome assemblages of bones thought to be secluded dining rooms for Sasquatch. Chronically barking dogs are often the ones that turn up missing in rural neighborhoods. It is thought that if a dog interferes with a nocturnal prowl route that the Sasquatch will simply get rid of it. And likely eat it too. One witness reported to the police that he saw his beloved and aggressive German Shepherd come flying over a tree into his yard, dead. It was a throw of at least 35 feet (10 meters).

In the information Alley received from witnesses, throwing of objects (including rocks, branches, and logs) is done underhanded or side-armed, not overhand.[332] The accuracy and distance of these throws are impressive and frequently way beyond the ability of H. sapiens such as the unfortunate, thrown German Shepherd.

Bears, particularly large grizzly bears, may receive careful treatment from Skanicum. Baxter writes:

> I have a theory that bear and Sasquatch do their best to avoid each other. I'm sure the occasional dispute occurs, but I think they both do what they can to avoid each other.[333]

He backs up his theory with an encounter from J. Robert Alley's *Raincoast Sasquatch* in which a grizzly bear and a Sasquatch—seen separately from a boat off shore—ran in opposite directions after merely catching the scent of each other.[334] Alley goes on to theorize that adult Sasquatch may have

---

[332] Alley 2003 (2018 edition), page 295

[333] Baxter 2021, page 85

[334] Alley 2003 (2018 edition), pages 91-92

no fear of black bears, but that they likely avoid coastal brown bears.[335]

Behaviorally speaking, Sasquatch seems to use knocking over or breaking trees as one of their ways of communicating. Sometimes perhaps, this mischief is simply a way of making their presence known. Sometimes it seems to be a warning. And sometimes pulling over trees seems to be done in anger. Wiitala writes:

> Speaking of pushing trees down, there are many different reasons for them to push them down. They might be mad, or just saying hello, or attempting to guide me, or for marking something.[336]

In the summer of 2025, I heard a tree break at a distance with a sound almost like a high-caliber rifle shot. Then came a great crashing of branches through other trees as the unfortunate timber slowly and then more quickly toppled towards the earth. Finally there was a boom as the trunk hit the ground. The whole process took an estimated three seconds. The noises were impressive and had my full attention, although the toppling was far from me and I could not see a bit of it. If I had reason to believe that the Boreal Bigfoot was involved (and I was lingering then at a spot in the woods where I had left apples for… *whomever*), or acting out of anger towards me, I would undoubtedly have found the crashing noises frightening.

Sasquatch activity is associated with a variety of sounds that all seem to be intended to communicate, though the message is often ambiguous. Grossinger summarizes the types of sounds that his witnesses in northwest Canada report as including loud shrieks, whoops, roaring screams, ear-piercing cries, and whimpers.[337] Larry Baxter, in his book *Abandoned: The History and Horror of Port Chatham, Alaska*, describes a personal encounter in which he and his wife heard deep thumping

---

[335] Ibid., page 234
[336] Wiitala 2021 page 55
[337] Grossinger 2022, page 18

sounds.[338] One of Isdahl's witnesses reported hearing a roar in Florida that sounded like a lion.[339] Frank reports hearing barking and laughter from a young female.[340] Wiitala recounted from personal experience the repeated angry screams of a large male Sasquatch in the darkness rapidly approaching his small group from what was initially a great distance. You will find a YouTube link to audio of the March 2013 Wiitala incident in the footnote below.[341]

Many witnesses insist that they hear mumbling, phrases, and unintelligible language and conversation coming out of forest.

Perhaps you have heard the powerful and intimidating sounds that a lion or tiger can make with their relatively enormous lung capacity. I have worked with 1,500-pound cows and steers and the energetic noises that they make merely when they are overdue for hay are impressive. Alley assigned a top end of 120 decibels for a Sasquatch roar, equivalent to sirens or jackhammers.[342] A 1,000-pound Sasquatch just saying hello should be capable of scaring the hell out of someone who is not prepared for the volume and the power behind the greeting.

One dark night in western Massachusetts, I was checking one last time on the cows and the chickens, when a high-toned, single syllable "**Who!**" came at me from about 75' away in the darkness and about 10' off the ground. I told my nervous self that I had simply heard a harmless little owl with anomalous decibel capacity. The sound was similar to the repeating "who" call of the local long-eared owl, but at a much, much higher volume and there was only the one syllable—no repeating call, as is their tendency. I couldn't see anything by the light of my feeble low-battery headlamp. The air felt electric as I assured the darkness that I was just making my rounds before going inside for the night. I still don't know what made that sound, but I believe I can eliminate an owl from the list of candidates.

Sometimes Sasquatch makes sounds that seem like the snorts

---

[338] Baxter 2021, pages 3-4
[339] Isdahl 2021, page 155
[340] Frank 2021, page 102
[341] Wiitala 2021 page 34; https://www.youtube.com/watch?v=xj99nrT2-84
[342] Alley 2003 (2018 edition), page 300

of a pig, as noted by Lapseritis[343] and, out of Alaska, from the September 9, 2013 Denali National Park & Preserve's *Wildlife Report*. That document had an *apparently* tongue-in-cheek paragraph involving a local ATV tour company operating just north of the Park. According to this entry in the *Wildlife Report*:

> …flustered and excited ATV riders have come out of the area mumbling words such as Sasquatch and Yeti. One even described a "man-bear-pig" before being hushed and moved off by the others.

A "man-bear-pig" is a very specific description to include in a tasteless "joke." I suspect that description originated in an *actual* report to disbelieving National Park Service staff.

Whistles are also commonly reported. Many North American native peoples warn that you should not whistle in the woods if you don't want eight-foot tall company. According to Short, the northern Hare branch of the Athabascans in northern Canada believed Bushmen lived in caves or underground, communicated with whistles, and used to eat men and steal women and children.[344] She also wrote that the Gwich'in of the Bonnet Plume River area of Yukon Territory believed that the local "Black Giants" would communicate with whistles.[345]

Lapseritis credits Sasquatch with high-pitched whistling[346] and Julie Scott in her 2010 book, *Visits From the Forest People: An Eyewitness Report of Extended Encounters With Bigfoot*, refers to quiet whistling and grunts.[347] In Isdahl's book (a source of much direct witness testimony), an east Texas man describes his history in the woods experiencing whistling, thrown pine cones, saplings woven together or snapped, and calls.[348]

We don't have to look further than to our own peoples to get inspiration for how Skanicum might be using whistles. Whistling

---

[343] Lapseritis 1998, page 19
[344] Short 2024 pages 105-106
[345] Ibid., pages 106-107
[346] Lapseritis 1998, page 19
[347] Scott 2010, page 73
[348] Isdahl 2020, page 280

tends to be used by tribal people that live or traverse far apart, in mountains, or in heavily wooded areas. It is also commonly used while hunting, even among Anglos in North America, as a way to keep track of companions without scaring game. Whistling may just be a way to get attention or to locate others, but it has also been used to transmit complex messages in war (in World War II Papua New Guinea) or to send messages to lovers (among the Hmong of China). A tribe on the Canary Islands has a whistling language that so effectively passes as bird song that birds will imitate the messages.[349] For beings hunting and moving through deep woods, whistling is a natural and helpful skill to use. Whistling skills seems to be particularly advanced in Sasquatch. Jorg Totsgi, editor of *The Real American* wrote (in 1924) that Sasquatch/Seeahtlk can imitate the song of any bird, especially jays.[350]

I was once hiking with a companion on one of the few established and frequently used trails in Denali National Park in the state of Alaska—the Triple Lakes Trail. We were close to the start of the northern end of the hike, working our way south and up a few switchbacks through thick white spruce. Below us, almost all the way down into the brushy creek bottom, someone started clumsily imitating a train whistle. It was possible at times to hear train whistles from that very location, being only half a mile or so from train tracks, but there were no trains going by at the moment. The train whistler was just taking it upon themselves to make train noises, randomly and poorly, near a large creek, out in the woods. The noise was peculiar enough for me to stop and listen. The noise maker wasn't actually whistling, but making a "wooooo-hooooo!" type of noise. The sounds did not even seem to be coming from the stretch of trail below us, but from even further down the hill, off trail. I concluded then that there must have been kids near the trail, goofing around. Later on, after we had hiked out to the lakes and back again, I realized that we did not see a single child on the trail all that day. Eventually, I decided

---

[349] Taylor 2018, https://listverse.com/2018/08/24/10-tribes-with-superpowers-you-wish-you-had/ [accessed 12/21/2025]

[350] Pyle 1995, 2017 edition, pages 154-155

it *was* a kid trying to make noises like a train whistle, but I am not convinced that the kid belonged to *H. sapiens*.

Other sounds credited to Sasquatch include the loud and sharp percussive clunks of assumed “wood knocks” and the more subtle clacking of rocks clicked together.[351] Isdahl had one witness in Washington state describe tree knocks loud enough to wake him up.[352] It should be noted that there is little in the way of eye witness corroboration of what is making these sounds and how they are made. Morehead, actively researching since 1971, agrees: “...to my knowledge, no one has ever seen a Bigfoot hitting a tree.”[353]

Scott reports whoops, whistles, grunts, growls, howls, screams, monkey-type chatter, clapping sounds (like they are made from banging together rocks or sticks), and an ambulance siren![354] Grossinger reports from the Yukon: “...verbal threats in the forms of yells and shrieks.”[355] Of the frequently reported barking sound, Grossinger describes it as “...similar to dogs barking but… much louder, guttural, and seemed to originate from a powerful chest, a sound that no dogs could produce.”[356]

A theory gaining ground about the “rock clacking”—which I myself have heard from about 100-feet away—is that it is made orally as a type of communication, perhaps by pulling the tongue off the roof of the mouth and “clacking” it against the bottom of the mouth.

I twice heard in one day the more rarely reported “tongue clacking” (which may be the same thing as rock clacking). The volume of both tongue clacks I heard that day (the first time from only about 80’) was more than I can match.

There have been a few reports of mother Sasquatch making a “weird low humming sound" to their infants.[357]

Some of the sounds associated with Sasquatch are bizarre and

---

351 Grossinger 2022, page 18
352 Isdahl 2021, page 150
353 Morehead 2024, page 113
354 Scott 2010, pages 110-111
355 Grossinger 2022, page 128:
356 Ibid., page 217
357 Isdahl 2021, page 145

inexplicable. More than one source, including Wiitala, has described a "metallic sound."[358] Twice while working on a western Massachusetts farm in the Spring of 2025, I heard a dirt bike on the heavily-wooded hillside above. The first time, I and another ear witness investigated and found no dirt bike tracks in the soft forest floor. We could not figure out how a dirt bike could even get around on that hillside with so many fallen trees on the ground. The other witness then thought that he could hear quiet voices in the woods above us which I could not hear at all. The second time that I heard the dirt bike sounds I didn't even bother to go look for the trespasser.

Wiitala thoroughly describes a rare encounter with Sasquatch sounds in southwest Washington state:

> ...I began to hear a juvenile male, a few hundred yards away, chanting on top of a ridge directly above. I really have a hard time describing the chant. It was a blend of four different stylized animal sounds. In other words, not a direct mimic of any animal. The four sounded like a grouse, cat, elk, and crow, maybe? He was very loud!.... Soon he was in my campsite [in complete darkness] chanting very loud and moving in rhythm to his chant. A literal sasquatch chant and dance. I could tell by listening where he moved and danced around...[359]

Another consistent behavioral characteristic of Sasquatch is their fascination with humans. They find us at least as interesting as we find them. They spend a great deal of their time and energy watching us in what seems to be pure curiosity for its own sake. Frank says he has been followed by the young female of the barking and laughing sounds and even tossed a stick back and forth with the girl Sasquatch, though she remained hidden from his view.[360]

Much of the 1924 Ostman account describes the curiosity of Ostman's captors: "...but the boy and the girl were always

---

[358] Wiitala 2021 page 41
[359] Ibid., page 59
[360] Frank 2021, pages 101-102

watching me from behind some juniper bush." In a longer passage that also demonstrates the apparent use of language, Ostman said:

> The young fellow was coming nearer me, and seemed curious about me. My one snuff box was empty, so I rolled it towards him. When he saw it coming, he sprang up quick as a cat, and grabbed it. He went over to his sister and showed her. They found out how to open and close it—they spent a long time playing with it—then he trotted over to the old man and showed him. They had a long chatter.[361]

Paulides refers to multiple reports of northern California Bigfoot watching Native American ceremonies and dances from the edges of the woods in the old days.[362] Scott points out that large windows with no curtains seem to provide a kind of television for the Night People, through which they are sometimes spotted to the horror of the residents—peering curiously through the glass back at them.[363]

My one sighting was from about 150' away in gloomy woods in Illinois in 2009 or 2010, but has the stamp of curiosity all over it (its curiosity *and* my curiosity). I did not perceive this incident as a literal Sasquatch sighting for *years*. On this particular day of an organized spiritual retreat, I had taken a new serpentine route through an unfamiliar grove on the retreat grounds. I was looking for a large pond that I had visited the day before. Deep in thought, it took me a few moments to become aware of a normally-proportioned dark figure downhill and a couple hundred feet away, moving stealthily through an area choked with brush and downed timber. I couldn't pick out any features, clothing colors, or transitions from shirt to pants, shirt to hands, etcetera. What I saw was just a black silhouette. Mid-day light was murky in the space under the deciduous trees, but I should have picked out some color transitions. The figure was slim, appeared slightly fuzzy around the edges, and stood perhaps five and a half feet tall.

---

[361] Bigfoot Research Organization (BFRO) Report #1091 2000; https://www.bfro.net/GDB/show_report.asp?id=1091

[362] Paulides 2017, page 473

[363] Scott 2010, page 9

The being stepped stealthily through the brush with absolute silence, attention focused intently on the sound of the murmured voices of a chatting group of human walkers on an established trail approaching from a different direction than my own heading. The figure that I saw seemed unaware of me, but it was *very* aware of the approaching hikers. It was high-stepping carefully over branches and brush and moving its head back and forth to spy on the humans around intervening tree trunks and branches. Then it appeared to pass behind a large tree trunk, but never came out from behind the tree again. "Where did it go?" I wondered. The whole sighting lasted at most ten seconds.

Though I had already been interested in Sasquatch for over thirty years at the time of the sighting, it wasn't until years later that I *finally* put it all together: Uniform black; fuzzy silhouette (hair-covered!); curious of, but hiding from humans—that was a Sasquatch! The reasons I took so long to come to that obvious conclusion included the small size of the figure and the location. My sighting occurred in a rural countryside of woodlots, subdivisions, and cornfields, but it was also only 40 miles from downtown Chicago, and just two miles from where the exterior scenes of the movie, "Groundhog Day" were filmed.

Sasquatch may also be curious about human *stuff*. In deep woods, along the foot-trail route to my off-road cabin in Alaska, there is a tall, white spruce tree. Under that tree, in about 2017, I found a small collection of kitchen ware, a Planter's Peanuts tin, and a coil of light gauge aircraft cable. This scattering of beat-up goods was well-removed from any cabins or human paths outside of my own. I never thought too much about it until Baxter reported finding a cache or hoard of modern trash in the woods in an Alaskan town abandoned in the 1950s.[364] Are there kids in the woods that like to collect human cast-offs?

More than one witness describes swaying back and forth,[365] likely a sign of nervousness or tension (in common with some Great Apes). In a *Flash of Beauty* video episode a witness describes how he saw by moonlight a large male Sasquatch step

---

[364] Baxter 2021, pages 76-78; photo page 115

[365] Grossinger 2022, page 84

into the trail in front of him. It was one hundred feet away or so, and began rocking back and forth, shifting its weight from foot to foot. The witness nervously pointed-out the swaying to his guide (a woman well-known in the Sasquatch community). She replied, "Then you should sway too." I don't know if joining the swaying was good advice or not, but the big male soon stopped swaying after the man started imitating the movement.

Lapseritis believes Sasquatch does not care for humans under the influence of alcohol[366] and Wiitala seems confident when he writes, "I've learned that they don't like alcohol, so I minimize that when I'm out there."[367]

Sasquatch has a reputation for being very strong. Fahrenbach explains why this makes perfect sense without resorting to any extreme reasoning. He points out that "…maximum *human* weight-lifting ability, extrapolated to the weight of the average Sasquatch (299 kg): it yields 1,300 lb. (610 kg). …its real capability is apt to be much greater than that of humans…"[368]

Once people start to accept that there is something tangible out there (*sometimes*), making noises and leaving tracks (*sometimes*), they naturally want to know if this creature's presence may be a danger to themselves or their loved ones. The answer is, "Probably not." However, there is a slim risk of engaging with something so big and so intelligent. The Night People adults (which I believe are seldom seen) likely want nothing to do with us puny humans, but the kids find us a source of great entertainment. And some of those kids are teenage boys that may be almost as inclined to create mischief as modern human teenage boys. Think about the potential there—teenage boys, with God knows what hormonal changes going on—six, seven, and eight feet tall, running around in the woods at night, looking in windows, and at a loss for something fun to do.

Even a sympathetic experiencer such as Scott acknowledges that the Forest People are generally harmless but, "…there are indeed some dangerous Bigfoot who can and do kill people."[369]

---

[366] Lapseritis 1998, page 45
[367] Wiitala 2021 page 96
[368] Fahrenbach 1997-1998
[369] Scott 2010, page 23

And she acknowledged that they kept a constant watch on their granddaughter when she was outside their hotspot home.

It is remarkable that there are so few reports of harm linked to Sasquatch. Consider the whole pesky researcher-in-the-woods-at-night situation. Those researchers must be *particularly* annoying to any Sasquatch trying to hunt and gather, but I only recall a couple incidents of people getting minor injuries. Our marijuana gatherer in the corn field had the most violent encounter. In another case from New Mexico, a rock took some bad bounces and cut someone's head. In yet another case, two teenage human boys got into a rock fight with what they thought were a couple of regular kids in the bushes. They were soon getting a whupping and saw that the four-and-half-foot-tall kids were hairy and naked. Those hairy kids were barely past the toddler stage and human teenage boys could not best them in a rock fight!

Of course, people do disappear in the woods. We don't know if Sasquatch has anything to do with those disappearances. There is more, known and unknown, in the woods besides Sasquatch. It's clear though that if Bigfoot wanted to mess us up, they could do us great harm. They could simply break every window in a house in about one minute and we would go screaming for our cars to get out of there. They actually show *remarkable* restraint considering the number of people still out in the woods with rifles looking to be the one who proves that Sasquatch is real. Pyle put it nicely when he wrote: "...everything we know about the American Yeti tells us that if it exists, it is infinitely more peaceful and non-violent than the other great ape occupying the continent." The "other great ape" he refers to is, of course, us: *Homo sapiens*. *We* are the dangerous and primitive species. Patrick found himself trapped in our primitive and dangerous world.

Incorporating all of these observed Skanicum behaviors, we may begin to answer Guermer's question, "What do you think Sasquatch is?" The very existence of the hybrid Patrick is another enormous clue into the nature of this enigmatic creature of the night. In the next chapter, we will look at more aspects of Patrick that he left for us to consider.

CHAPTER SIXTEEN

Patrick's Physiology

"The general shape of the head is that of an ellipsoid or an ovoid, and its average height is 18–20 cm. **This measurement is usually relatively constant as 13% of a person's total body height,** although it does vary somewhat with ethnicity, sex, and age. The head is comprised of two subdivisions, the cranium and the face. The cranium contains the brain**…"**

Makris, Angelone, et al. 2008, *MRI-based anatomical model of the human head for specific absorption rate mapping*, page 2

I left the Patrick story at the end of Chapter Fourteen with an eleventh hour discovery of an arrest record from October 10th, 1928. I hinted at something remarkable associated with that arrest record. This chapter is the big reveal.

Patrick was arrested in Kittitas County, Washington where the largest city is Ellensburg, but the county also contains Cle Elum and Roslyn. Patrick may have gone to Kittitas to pick hops or fruit or for some other cash labor. His crime of arrest is not given. He was sentenced to 30 days and $150 (worth about $3,000 in

2025).[370]

Only two people were arrested that day: Patrick and Pete Anderson, a "Sweed." The arrests immediately before Patrick and Pete occurred on 10/08 (two arrests) and those immediately after Patrick and Pete were arrested did not happen until 10/13/28 (four arrests). It is gross speculation, but Patrick and Pete may have been arrested *together*, as drinking buddies, or perhaps as the result of a dispute. If it was a dispute, Patrick chose the wrong man: Pete is described as 6'-0" and 190 pounds to Patrick's *overstated* height of 5'-7" and likely accurate weight of 135 pounds. There is a tenuous piece of evidence leaning me towards this arrest having involved a dispute with blows. Looking closely at Patrick's remarkable *mug shot*, there may be swelling and discoloration to Patrick's lower left chin and mouth area corresponding to someone connecting with a right-hand hook. Or I could be seeing traces of an unrelated injury noted on his WWII Draft Registration card almost 14 years later: "Scar on chin".[371]

This arrest record again indicates that Patrick was born in Chelan, was 38 years old (I think he was actually 35), his complexion "red" (shorthand among whites of the time for Native American), his eyes brown, and his hair black. In his mug shot, Patrick is wearing a light, high-collared shirt, overalls, and a sweater or light jacket.

Pete the Swede received the same jail time of 30 days, but apparently was not levied a fine.

Patrick is shown in two pictures that are almost a century old at the time of this writing. These are the only photographs that I have found of Patrick. I am pretty certain that a couple more pictures were taken of him 22-years later in 1950, but I have failed to find them after a long search. There may also be old family pictures out there somewhere still, moldering in an attic box.

At first glance, he doesn't look all that odd for a guy who was fathered by a Sasquatch. He has pretty much normal hair patterns. No sign of a beard or moustache. No conical skull and no heavily-

---

[370] October 28, 1928, Patrick [redacted]; Mug Shots; Kittitas County; Washington State Corrections & Jail Records

[371] April 27, 1942, Patrick [redacted]; Registration Card; page 2

muscled neck. His brows aren't unusually protruding as far as we can see in the front view or in the poor profile picture. The nose looks to fall within the wide range we find in *H. sapiens*.

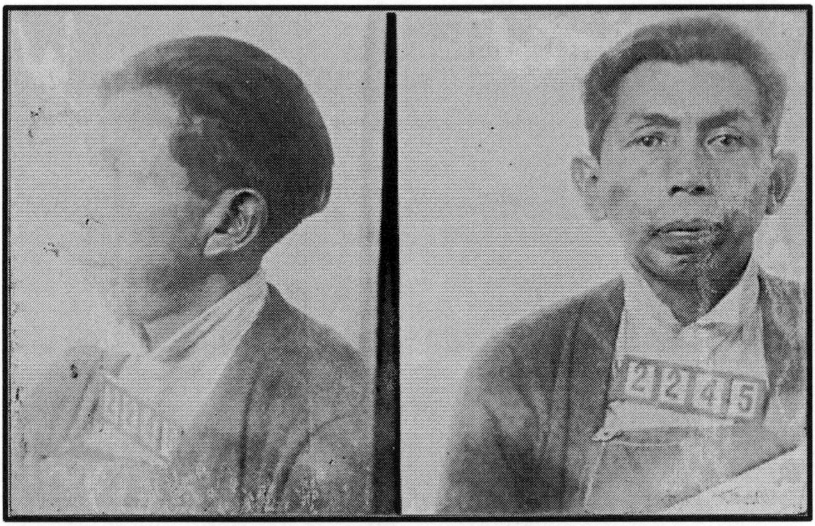

Patrick, the Sasquatch/human hybrid pictured from 10/10/28 when he was about 35 years old in Kittitas County, Washington state.

If you linger on the picture though, you may find his overall appearance peculiar. Looking closer, some oddities include very full lips; a lot of space between the bottom of his septum (nose) and the top of his upper lip. Or is that impression simply due to a weak chin? It's hard to tell, but his skull seems to be widest at his cheekbones (not uncommon in Native Americans). And artists are told that the eyes should be centered about halfway between the top of the skull and the bottom of the chin. But these eyes are higher up the head than that. In the profile picture it may look to you like Patrick's skull extends a bit more rearward than normal. And it's tough to see in the profile picture, but his forehead may slope a bit more steeply backwards than the norm.

Finally, look at the ears. Patrick's ears are truly odd. Look at yourself in a mirror and you will find that an imaginary horizontal line running through both of your pupils, and extended to each side, appears to pass through the tops of your ears. Patrick's ears fall entirely *below* that pupil line. Additionally, the bottom of modern human noses are typically in the plane of the bottom of the earlobes. Patrick's ears are atypical in respect to both of these markers. They are *extraordinarily* low on his head.

Unsurprisingly, this ear arrangement is known as "low-set ears" and may be associated with several congenital (from birth) syndromes that all tend to be linked to chromosomal abnormalities. Of these syndromes, two are the closest matches to what we can see in Patrick's images: Edwards syndrome (also known as Trisomy 18) and Noonan syndrome.

Edwards syndrome is characterized by low-set ears, small birth size, small head, and small jaw. However, Edwards syndrome is only rarely inherited, is more common in older mothers, and usually involves low IQ.

More likely in Patrick's case perhaps is Noonan syndrome. Among effects in common with Patrick include small lower jaw, low-set and rearward-rotated ears, and a short and webbed neck. However, Noonan's syndrome is normally also accompanied by congenital heart disease, abnormal ear shape, drooping eyelids, an upturned nose, and eyebrows slanting to the outside of the head. Noonan's syndrome describes some of Patrick's features, but can't be characterized as a good match.

You may have read the story of Zana of Abkhazia (a region of the southern Caucasus, near Georgia and eastern Turkey) a supposed tall, muscular, and hair-covered wild-woman or *Almas* held in captivity and bearing six children by local men. Burtsev published pictures of some of the descendants of Zana (two on the next page). Note the ear placement on these two relatives of Zana (see the photo on the next page)—atypical once again!

Does ear placement on the unidentified woman and Zana's son, Khwit, imply a link or connection to Patrick?

On the face of it (pardon the pun), *no*, there apparently is no link between Zana and Patrick. Margaryan et al produced a thoughtful paper in 2021: *The genomic origin of Zana of Abkhazia* in which they describe their successful efforts to determine Zana's genetic heritage. They found that Zana was of primarily East African descent with perhaps a minority West African admixture. They suggest that Zana was a victim of the slave trade from East Africa to Turkey and suffered from congenital defects including congenital generalized hypertrichosis (CGH) or extreme hairiness.

Genetically speaking, it is impossible for me to argue with their results. And a co-author of this 2021 paper was none other than Igor Burtsev![372]

From Burtsev 2015, pages 104-105: On left: "Zana's daughter or granddaughter, not identified" and on right: "Zana's son Khwit." Note on both the atypical placement of the ears, both the tops of the ears below the plane of the pupils and earlobes falling well below the bottom of the nose. Unusual ear placement is visible in pictures of Zana's great-grandchildren.

So why do I bring Zana up? Zana's description does not seem satisfactorily explained by this proof of sub-Saharan African origin and speculation of CGH. Zana was described as 2 meters tall (6'-6"), very strong, fast as a race horse, and covered all over with long reddish hair. It doesn't seem likely that a slave woman suffering from congenital defects and found naked in the woods would be tall, strong, and fast. And would the effect of CGH on a sub-Saharan African produce long reddish hair?

One last thing that makes me hesitate before writing Zana off: her mitochondrial DNA occupied a branch of the L2 haplogroup (L2b1b). The same haplogroup as found in sample 11 (L2c) in the Ketchum et al study and identified as originating in Africa.[373]

There will be more on these inconsistencies and Zana in Volume II.

Much hard anatomical data can be extracted from these

[372] Margaryan et al 2021
[373] Ketchum et al 2012, page 15

pictures (of Patrick) and examined to see if there is anything truly odd about Patrick's face and head. You may recall the section about Patrick's appearance from Ed Fusch's paper. Louie, another Moses-Columbia Indian and 79 years old according to Ed when he interviewed him about 1985, provided some interesting recollections about Patrick's appearance. Louie would have been born around 1906 and was about 19 when he started working for Patrick in approximately 1925. According to Louie's recollections, he worked with Patrick until about 1930, two years *after* these mug shots were taken. According to Ed, Louie "describes him as a 'pinhead' about 5'4" tall, with larger than normal ears, mouth, teeth, and with large hands…"[374]

We will look at the pinhead claim soon. From what I have seen in other documents, I suspect the 5'-4" estimate was on the generous side. Perhaps surprisingly after all these years, I have found 13 official documents that record Patrick's height over a span of twenty years, beginning with this arrest on October 10, 1928. He is recorded once for each of the heights of 5'-0", 5'-1", and 5'-7". At no point was he recorded as 5'-3" or 5'-6". Four times he was documented as 5'-4" and most frequently, six times, he was noted to be 5'-2", including on his WWII Draft Registration card in 1942 when he would have been almost 50 years old.[375] I suspect that he tried to pass as 5'-4" and sometimes people took his word for it, but when he was actually measured he usually came up short at around 5'-2".

Therefore 5'-2" (1.57 m) is the measurement that I am going to use for the rest of this chapter and book.

I don't think these pictures agree with the assessment of large ears. And his mouth also does not seem wider than normal, but I have already noted the very full lips. We can't see what his teeth look like. And we will have to take Louie's word on the large hand observation.

In relation to the "pinhead" charge, remember what Makris, Angelone, et al wrote:

---

[374] Fusch 1992a, page 22

[375] April 27, 1942, Patrick [redacted]; Registration Card; page 2

> The general shape of the head is that of an ellipsoid or an ovoid, and its average height is 18–20 cm. This measurement is usually relatively constant as 13% of a person's total body height, although it does vary somewhat with ethnicity, sex, and age.[376]

So, with Patrick at 5'-2" (157 cm) tall, his head height should be about 13% of 157 centimeters:

= 0.13 x 157 cm
= 20.4 cm = 8.035"

The head height of an average heighted adult modern human male at 5'-9" (175 cm) tall would be 13% of 175 cm:

= 0.13 x 175 cm
= 22.8 cm = 8.97"

The math tells us that we should expect Patrick's head to be about 1 inch shorter than the average adult male of the United States today. Presuming eye width (from pupil to pupil) varies in proportion to height, Patrick's eye width would also vary by the same ratio as his height to the average height or:

157 cm/175 cm = 0.90

So we will multiply the soon-to-be-measured standard 5'-9" male's eye width by 0.90 to get Patrick's expected eye width (presumably approximately the eye width of any 5'-2" *H. sapiens*).

Figure 1 is of a contemporary Colville man's head. I used AutoCAD to make the Figures of this chapter and to measure features. A "normal" man's head, when scaled to an overall height of 5'-9" (175 cm), should have a total height of 8.97" (22.8 cm), as calculated above. In CAD-space, the distance between the centers of his pupils then measures as 2.66" (6.75 cm).

---

[376] Makris, Angelone, et al 2008, page 2

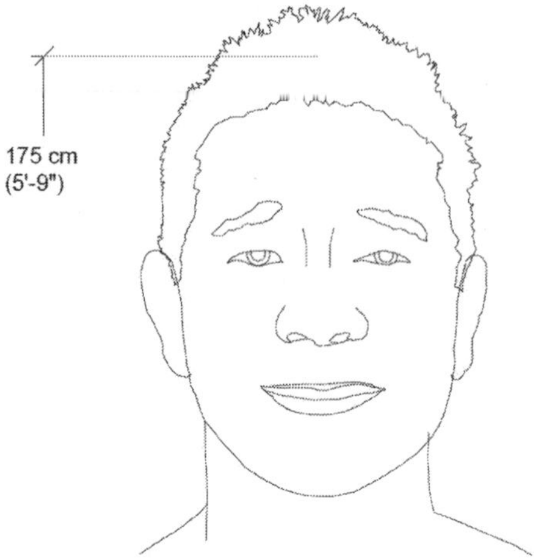

Figure 1

I am going to use this pupil-to-pupil width, rather than full head height to scale Patrick's head because I already believe Patrick's head likely does not follow that 13% rule, anticipating an "ethnic" variation in cranial structure. I am more confident that the distance between the center of his pupils followed the head height scaling. Using the ratio calculated above, we would expect Patrick's eye width to be 0.90 times that of the eye width on our 5'-9" Control, or:

Patrick's eye width = 0.90 x 6.75 cm
= 6.08 cm = 2.39"

Below is a tracing of Patrick's head from his October 10, 1928 mug shot with his pupil-to-pupil width scaled to 2.39" (6.08 cm): With the help of the AutoCAD program, I have placed Patrick at 5'-2" next to our Colville Control at 5'-9", each at the same scale, and with the addition of a suggested scalp profile (see Figure 2).

You may point out, "Sure, Patrick is smaller than a 5'-9" "normal" man, but we already know he is going to be smaller because he's *short*." Right you are. In a minute, we will scale the modern Colville Indian down to Patrick's 5'-2" height and see if that takes care of the differences that we are seeing. But one last thing to look at before we start comparing more along the lines of apples to apples: The relative length of Patrick's neck (see Figure 3).

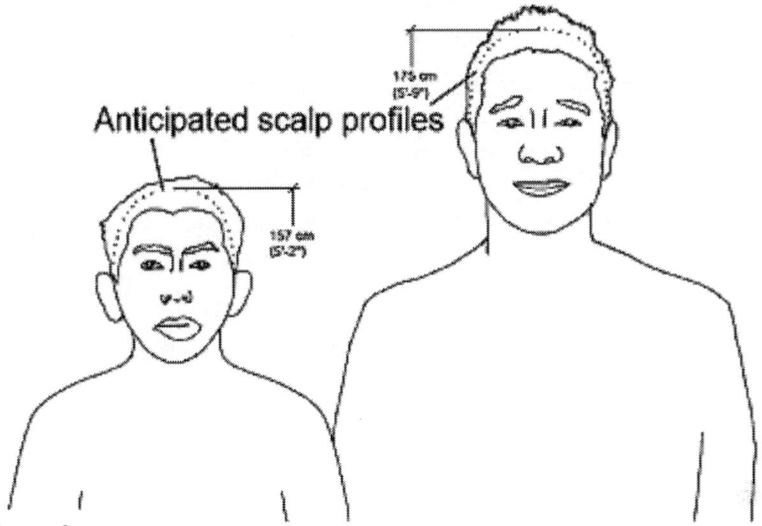

Figure 2

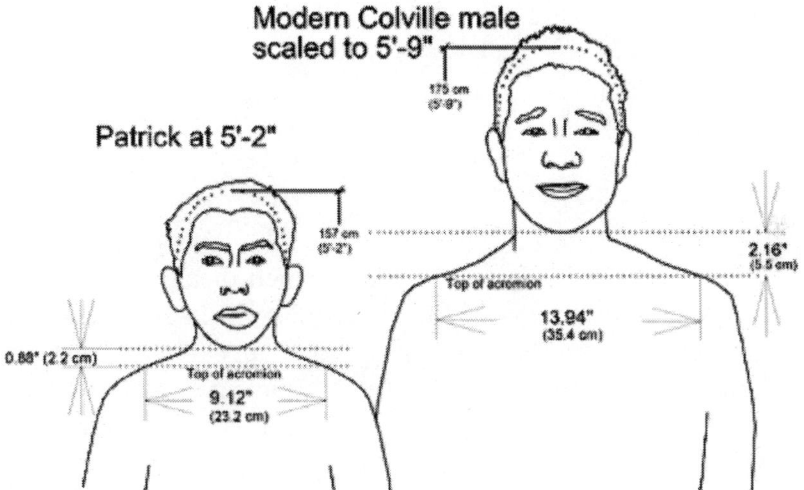

Figure 3

The vertical high point of the scapula bone's acromion process (a *bone process* is a projection or extension of the larger bone mass). This is the pokey-up bit that you can feel slightly front of the center of the top of your outer shoulder, generally immediately above the armpit. I am using the acromion here as a fairly identifiable landmark on the shoulders of Patrick and our

Control figure. The bottom of the chin (or lower jaw) is another easy landmark. The distance between these two landmarks should have a consistent correspondence with the relative length of the neck.

The distance between the top of the acromion plane and the underside of the chin on Patrick is less than half of the same measurement on the 5'-9" Control figure. If we scale the Control figure down to 5'-2", as in Figure 4, we still see indications that Patrick's neck is comparatively short:

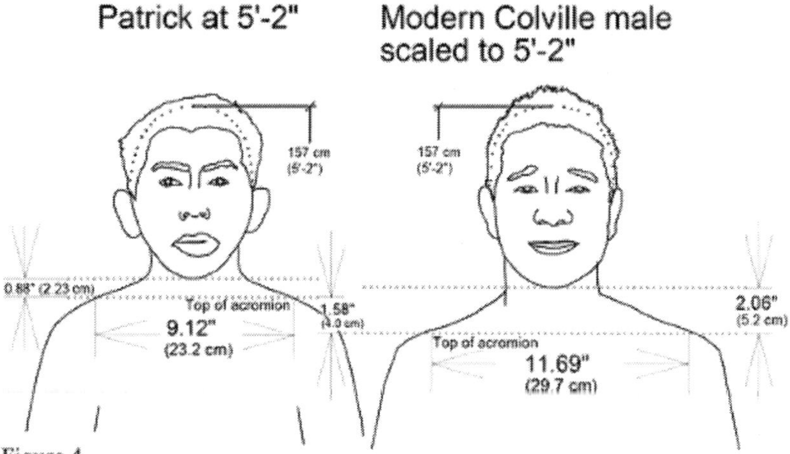

Figure 4

There appears to be something radically off about Patrick's neck, specifically the cervical spine—which dictates the neck's length. Just as in reports of Skanicum, Patrick has a noticeably short neck. But in Patrick's case, the appearance can't be explained away as due to heavily built-up neck and shoulder muscles. His neck is *structurally* short. This measurement as drawn—though not equivalent to the full length of his neck—is 57% less than that measurement on our Control *H. sapiens*, even when that Control is scaled down to Patrick's height.

In *H. sapiens*, the cervical spine is composed of seven vertebra. In Patrick's case these seven vertebrae are either compressed vertically, perhaps through injury, a degenerative effect, an inherited trait, or he was not born with all seven vertebrae. Interestingly, Neanderthals were found to have had vertically compressed cervical spines and shorter necks than *H.*

sapiens as discussed in the 2025 paper by Carlos A. Palancar, Daniel García-Martínez, and Markus Bastir: *The Neanderthal cervical spine revisited.*[377] In that paper, it is suggested that this "craniocaudal" shortening, coupled with increased curvatures of the upper spine, may be an adaption to support heavier heads. Patrick does not seem to have had a particularly heavy head, but perhaps his father did.

Along with the short neck, his shoulders, as indicated by the distance between his acromions, are also comparatively narrow—22% less than the same width on the scaled-down Control figure. This correlation between a short neck and narrow shoulders may not be causative, but perhaps there is a structural reason why a short neck might go along with narrow shoulders or vice versa. Perhaps the narrow shoulders were accompanied by a barrel chest—the acromions squeezing together in concert with the sternum, ribs, and collar bones pushing forward—resulting in a similar circumference and torso volume as that found in our "normal" Control. It should be noted that Patty's torso was barrel-shaped, as were those of Neanderthals. Unfortunately, it is difficult to tease good data about the shape of Patrick's torso from the mugshots.

This radically short neck and narrow shoulders are two of the significant findings from these 1928 mug shots and might represent physiological traits influenced by his father's genetics.

Before leaving the frontal comparison of the two figures scaled to the same height, note the potential for Patrick to have had a relatively long torso. The plane of his acromions is over 1-1/2" (4.0 cm) above that same feature of the Control's. Even if Patrick's legs were exactly the same length of those of the Control legs, his torso would be longer than the Control torso. I have no information on the length of Patrick's legs relative to that of his body, but it seems likely that his torso was relatively long compared to his leg length, in keeping with what Meldrum, ThinkerThunker, and many others have observed in Sasquatch.

Patrick's projected frontal head area is 26% less than what you might expect from an average heighted male in the United States

[377] Palancar, Garcia-Martinez, and Bastir 2025, page 6

these days (see Figure 5). Looking at it this way, Louie's "pinhead" label may not be unreasonable at all. However, we still are comparing apples to oranges.

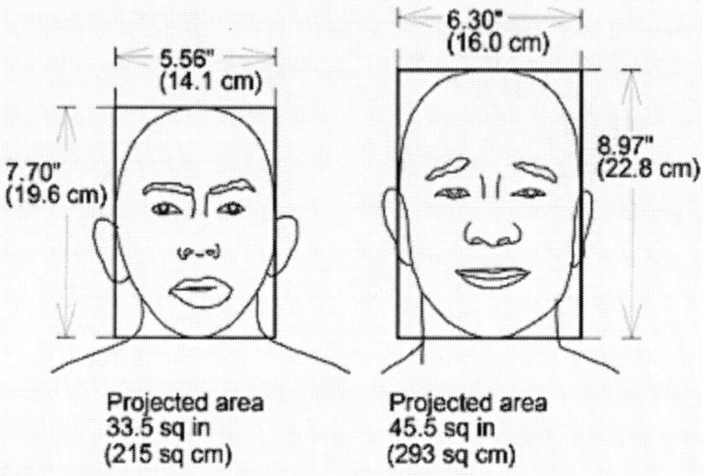

Figure 5

If we scale down the 5'-9" individual to Patrick's height, how does Patrick fair then? Figure 6 is the scaled frontal head profile comparing Patrick's head with a typical Colville Indian scaled to the same overall height as Patrick's assumed 5'-2":

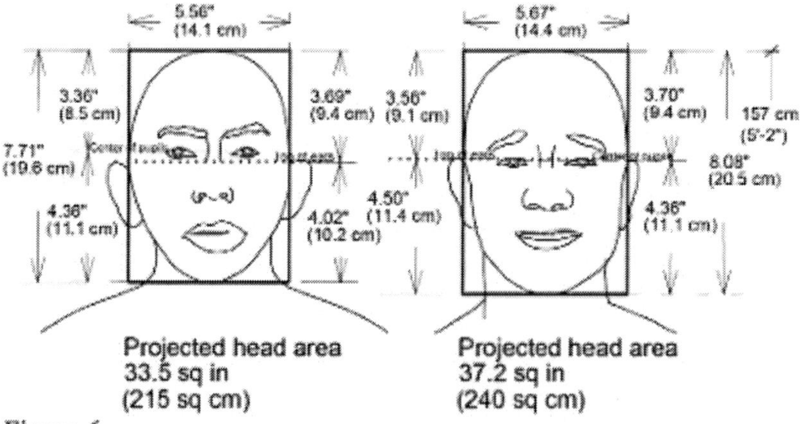

Figure 6

There are a couple features of note in Figure 6. Perhaps surprisingly in someone labelled a "pinhead", the overall head height is not obviously one of these features. Though Patrick's head height is slightly shorter than our Control's (7.71" versus 8.08"), it is still 95% of the expected head height. However, the frontal area of Patrick's head, measured in the flat plane, as shown above, is only 90% that of the Control's area. Since a change in area squared equates to the change in the volume [in this case: $(33.5/37.2)^2$], we may expect that this drop in Patrick's frontal area could indicate an approximate 18% drop in total head volume compared to our scaled Control. Does this estimated reduced head volume reflect a real drop in cranial volume? I will explore this possibility soon.

In this discussion of Patrick's physiology, I am not going to stress Patrick's anatomical differences if the percentage change from that of the Control is less than 10%. I am presuming that single-digit changes can be reasonably credited to normal variations among *H. sapiens*, such as by ethnicity, gender, and age as suggested by Makris, Angelone, et al. (2008) in this chapter's epigraph. A comparison of all of the measurements, including those in which Patrick is less than 10% removed from Control measurements, may be found in Table 2 at the end of this chapter.

Another peculiarity of Patrick's head mentioned earlier is the placement of his ears. Look at yourself in the mirror, or at your partner, your kids, at people in videos, at people at the coffee shop, or at pictures in a magazine… The tops of human ears are level with or above the eyes. This is how we are made and this is how artists are taught to place the ears in portraits. Because if you don't, the face looks *wrong*.

The plane of the top of Patrick's ears is *below* the center of his pupils. This is very unusual. It is not a feature of the angle of the mug shot. If anything, the camera lens is looking up at Patrick which would tend to make the ears look higher than they actually are. Go back to the side view in Figure 1 and check for yourself. As a friend remarked to me, "Of course the ears are a little higher than the eyes. We need them there to hang our glasses on."

Let's see what might stand out in side profiles. Note: The side profile mug shot is very poor, particularly when it comes to the front of Patrick's face. [The fuzziness of the image is interesting in light of how hard it is to get a clear picture of Sasquatch.] I had to monkey-around with picture settings, extrapolate, interpolate, and *guess* to arrive at the profiles below. This uncertainty of course then extends to the related measurements and Figures.

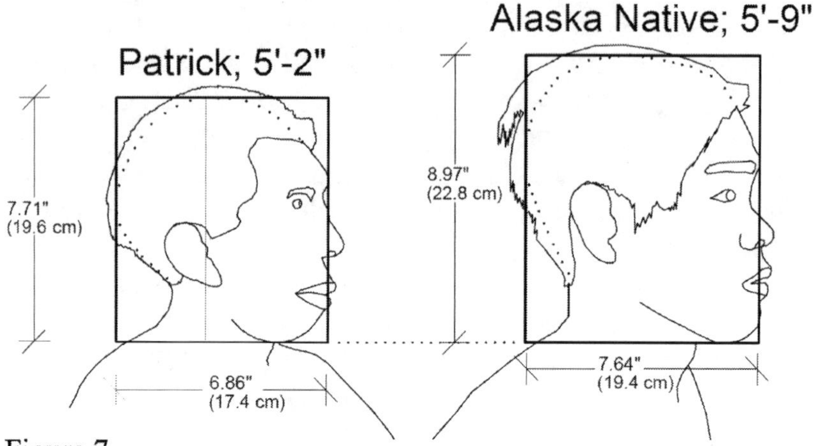

Figure 7

The Figure 7 view compared to the much taller "Control" male (in this case, an Alaskan Native from Patrick's era, scaled to 5'-9" tall) does make Patrick's head seem small. What if we scale the Control down to Patrick's height as in Figure 8?

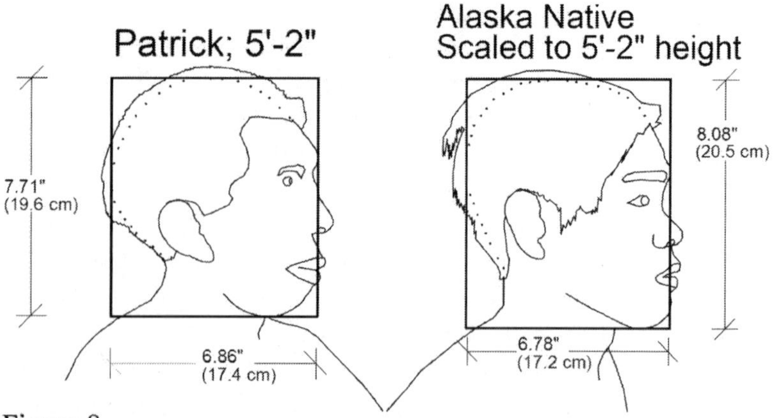

Figure 8

Patrick's head is still shorter than the scaled down Control head, but now we can see something interesting: His head is a little bit *deeper* from front to back than the scaled Control head. The effect is not striking, but may turn out to be significant in terms of relative brain volume.

The hint of a broken nose in the lump on his bridge may be entirely a poor image effect. I cannot vouch for the accuracy of that area and I have not found secondary evidence that his nose was ever broken.

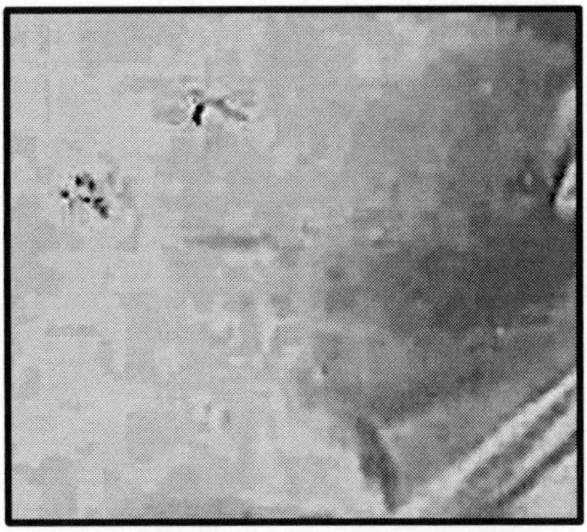

Potentially enormously significant: Patrick does not appear to have a chin in the anatomical sense of a forward-projecting bone at the front of the lower jaw (also called the *mentum*). Feel your own. Your chin may be somewhat concealed by flesh even in your side profile, but I think you will find it. I do not see this boney forward projection on Patrick. The side image of Patrick is poor, but I believe I have the chin profile very nearly correct. If Patrick indeed did not have a chin, this would be a *very* odd feature. As Schwartz and Tattersall wrote in their 2000 paper, *The human chin revisited: what is it and who has it?*: "We conclude that one can identify a chin, complete with detailed morphology, in all late Pleistocene fossils traditionally assigned to modern H. sapiens..." and "although there is variability within the Neanderthal sample in the presentation of the features described above, no adult Neanderthal specimen possesses, even vaguely or subtly, the inverted ''T'' and attendant mental fossae that are otherwise seen in H. sapiens."[378]

I do not see this detailed chin morphology in Patrick. Interestingly, the chin is an almost uniquely *H. sapiens* trait. There are a very few late Neanderthal jaws that show *some* chin (possibly a sign of genetic admixture with *H. Sapiens*), but generally speaking only *sapiens* have a mentum, even though there is no good argument for their functionality or necessity. For Patrick to not have a chin is puzzling except in the unique context of him having a Skanicum father.

---

[378] Schwartz & Tattersall 2000, pages 370, 383-386

This lack of a chin may have been yet another factor that gave Patrick the appearance of having a small head.

Let's look at some measurements of the side profiles with the hair removed to reveal anticipated scalp lines.

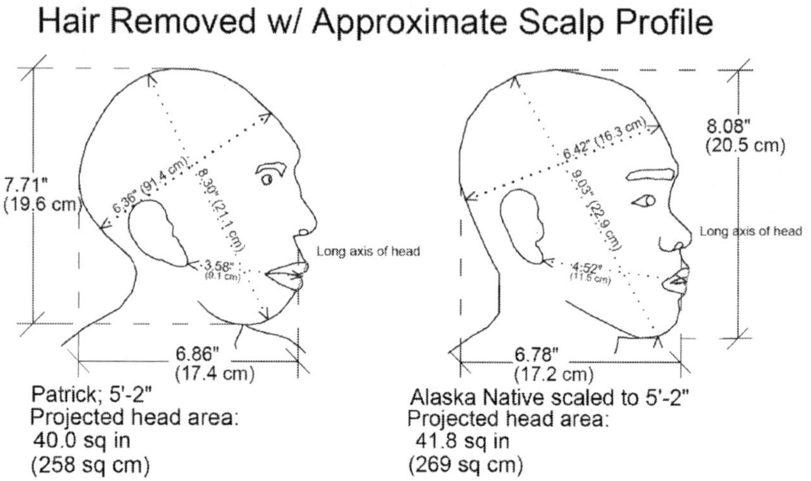

Figure 9

Some interesting points may be made from these profiles in Figure 9. Although Patrick's head is significantly shorter than the Control's head in the long axis, the length of the line at the circumference of his head (the sweat-band plane) is very nearly the same as that of the Control's (6.36"/91.4 cm versus 6.42"/16.3 cm). What really catches my eye though, is how much shorter the distance between his ear lobe and the corner of the mouth is on Patrick versus the Control's (3.58"/9.1 cm versus 4.52"/11.5 cm). It's as if the cranium (the cranium is that part of the skull that encloses the brain) was separated from the face along a line running from the bridge of the nose to just under the ear and rotated down and back from a pivot or hinge point in the temple close to the eye, thus lengthening the brow and radically reducing Patrick's facial area.

There is more information to be coaxed out of these side profiles. "Anterior" is a fancy way of indicating the front of an object. In Figure 10, I note the area projections of the front and rear sides of the heads. The anterior area of Patrick's side profile is significantly less than the same on the Control. Posterior area is less on Patrick as well, though the difference does not rise to the level of my arbitrary 10% cut-off. I believe that these differences in areas are due to his shorter head and more steeply sloping forehead.

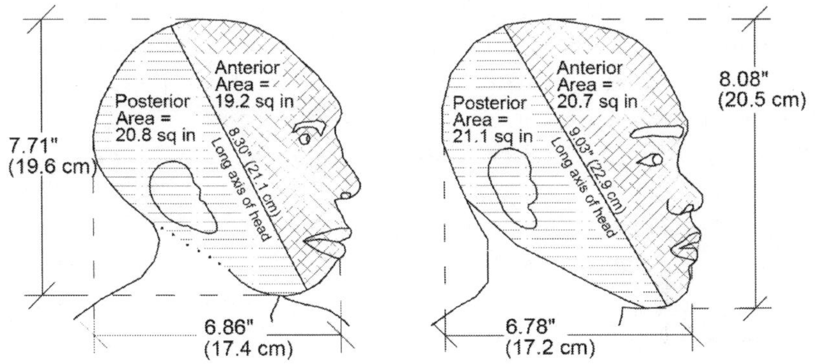

Figure 10

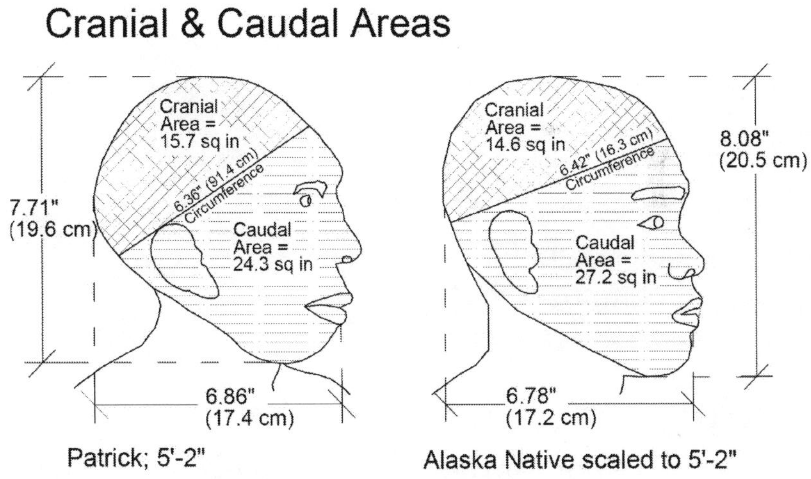

Figure 11

In anatomy, structures above another structure are sometimes referred to as "cranial," while the structures underneath are called, "caudal." In Figure 11, I somewhat arbitrarily used the circumference line established earlier to divide the side projections of the two heads into cranial and caudal areas, This division into top and bottom does seem to reveal some significant anatomical differences. Patrick does have more cranial area, though not significantly so, however Patrick's caudal area is 11% less than that of the Control's. This caudal area includes the face. Patrick's relatively low caudal area may have been what gave Louie the impression of "pinhead." However, the tables turn when we look at the ratio of projected cranial area to total projected side profile head area. The Control's ratio is 0.35, while Patrick boasts a ratio of 0.39 or 12% greater. From this we may begin to suspect that while Patrick was relatively small-faced, he may also simultaneously be able to claim being relatively big-brained! It is still early to make that claim though.

Let's put the two heads together in Figure 12, with the bottoms of the lower jaws at the same height in order to more clearly perceive any cranial differences:

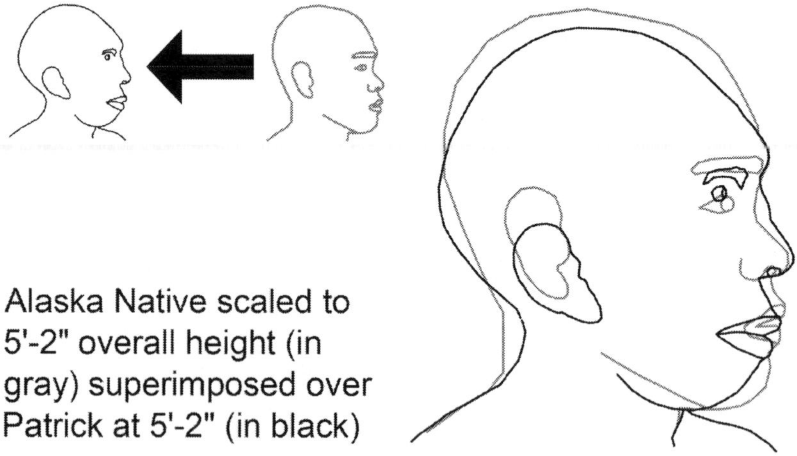

Figure 12

In this overlap, the "overhang" of the back of Patrick's skull is more evident. The extent of this cranial overhang is unusual, though not unprecedented in modern humans. And his ear really does seem to be remarkably tipped back, low-set, and pulled forward at the lobe. Seen in this context, with the corners of Patrick's lips so close to his ear lobes, one might see that Patrick could give the impression of having an unusually large mouth.

The vertical distance from the underside of Patrick's jaw to the top of his ear is 4.0" (10.2 cm), 10% less than the same measurement on the Control at 4.5" (11.4 cm).

Let's compare some of the angles of head features between the two subjects in Figure 13:

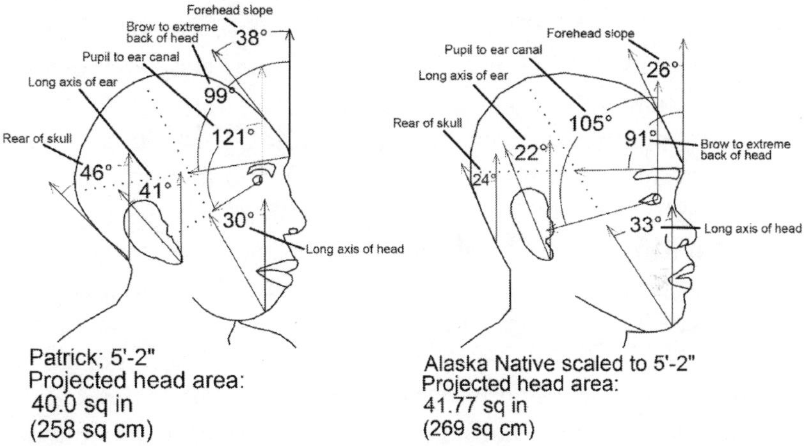

Figure 13

With these angles we find some startling deviations in Patrick's head structure from the Control's. His forehead slope from the vertical is a full 12 degrees or 46% greater than that of our Control figure (38° versus 26°). The line drawn from the pupil over the point of the ear canal is 16 degrees steeper from the vertical on Patrick (121° versus 105° = +15%). The long axis of the ear is tipped an extraordinary 19 further degrees on Patrick than on the Control (41° versus 22° = +86%!) and the rear base of his cranium is a full 22 degrees further from the vertical (+92%). It should also be noted as of potential significance that Patrick's skull is more elongated (dolichocephalic) than the Control's head in light of the maximum brow to back of head skin-surface measurement (17.6 versus 17.3 cm) compared to the maximum width measurements (14.1 versus 16.0 cm).

Patrick's head structure is *very* unusual.

The "extremely wide" and robust Willandra Lakes Hominid 50 (WLH 50) partial Pleistocene skull found in Australia fits fairly neatly inside of Patrick's side profile after considerable scaling down. I will devote more time to the WLH 50 skull in Volume II, but just mention here that this skull is exceptional in its unusual size, high cranial capacity (1,540 cubic centimeters versus the modern average of 1,300), sloping forehead, and *astoundingly* thick skull. Flood described the difference between it and other late Pleistocene skulls from Australia as the difference "between earthenware and bone China." Australian anatomist Alan Gordon Thorne described WLH 50 as "much more robust and archaic than any Australian Hominid found previously."[379]

Worth noting here may be that Patrick's cranium, though appearing small through its placement low and favoring the back of his head, does not have much resemblance to the supposed Yeren/Human cross that was mentioned in the abduction chapter. In the case of this other possible hybrid, the facial area appears much larger than the part of the skull devoted to gray matter (see the picture below). Note how odd the thumb and forefinger appear together. Although roughly parallel, the two digits extend about the same distance and their pads come together almost like tongs in a very un-human-like way.

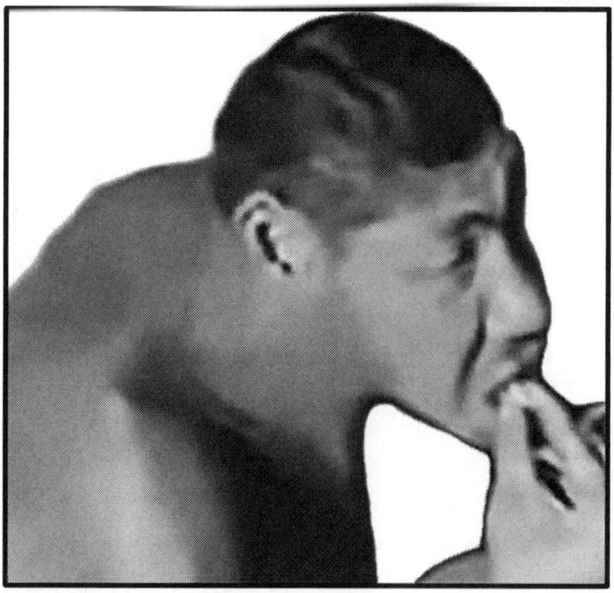

[379] Flood 2010 revised edition, pages 69-71

Finally, let's tackle the sensitive topic of Patrick's brain size. The brain is cushioned with a thin layer of cerebral-spinal fluid. The brain itself is, of course, smaller than the cranial void for this reason, resting in a protective waterbed of sorts. In producing the brain outlines below, I am using 0.59" (15 mm) as the average distance between the outer surface of the scalp and the surface of the brain per Hausinger, et al., 2011.[380] My amateur attempts at locating brain matter are bound by this average scalp separation distance and the typical brain location relationship with the ear canal and the eye and brow area. I acknowledge that much more accurate work on plotting Patrick's gray matter location, areas, and volumes can and should be done (see Appendix).

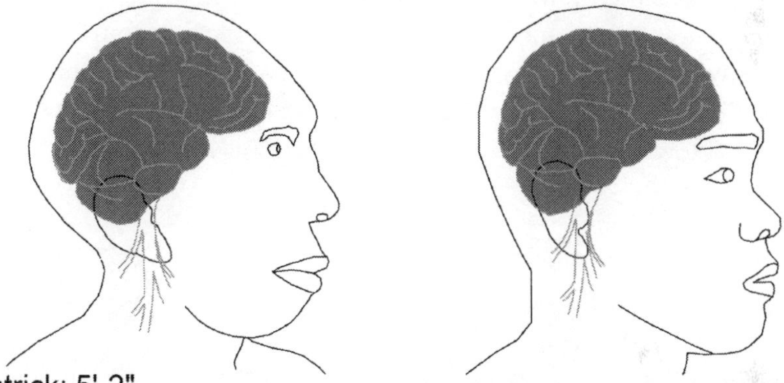

Patrick; 5'-2"
Projected cranial area:
14.97 sq in
(96.6 sq cm)

Alaska Native scaled to 5'-2"
Projected cranial area:
14.99 sq in
(96.7 sq cm)

Figure 14

Sidenote: Although the exact figure is somewhat disputed, most sources agree that the human brain *has shrunk* in the last 20,000 years by 10-20%. In 2014, Christopher Stringer responded to a letter to the editor of *Scientific American* asking if this troubling ongoing dwindling was in fact occurring. Dr. Stringer acknowledged this trend, citing possible causes of this reduction: Brain size, generally proportional to body size, may be decreasing as our bodies grow smaller on the average with a warming

---

[380] Hausinger, et al. 2011; page 5, figure 3 and page 6, figure 5

climate—or—*self-domestication* has reduced the need for big brains! Apparently the more ordered and predictable civilized environments we now enjoy do not require the processing power needed by our larger-brained foraging ancestors, so we have downsized. Both of these explanations imply that we might expect Skanicum, and Patrick, to have bigger brains than the average civilized human being.

Let's test this new hypothesis by looking at the side profiles of these two heads with estimated brain locations produced with my own tiny self-domesticated brain in Figure 14.

As I mentioned previously, the back of the brain is tied into or locked into relationship with the placement of the ear canal. Since Patrick's ear, for whatever reason, is lower and further forward, this presumably indicates that the volume of the brain has followed—it is occupying more space low and forward than in the Control with its higher ear placement. In the side view, the contest between projected cranial area is too close to call (only 0.1% difference).

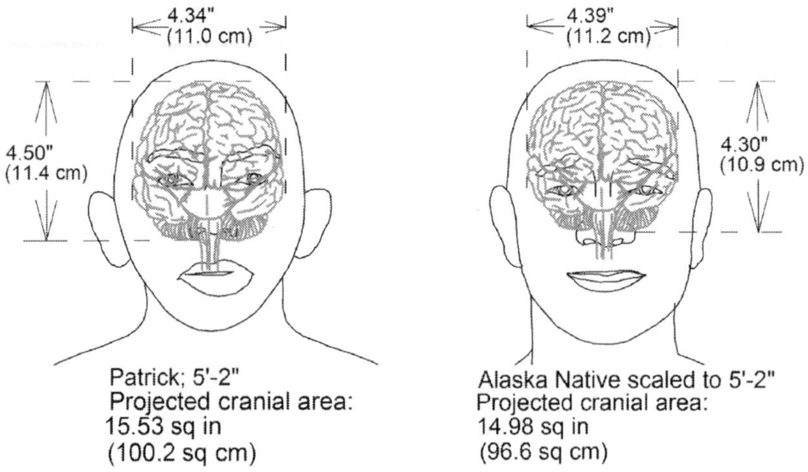

Figure 15

In the front view (Figure 15), Patrick's brain volume expansion with his unusual ear canal placement, results in a modest, but measurable gain of 3.7% over the Control's in terms of projected brain area.

This perceived "brain gain" may be an unintended product of

compounded uncertainties in reproducing Patrick's cranial anatomy, but it seems likely that even though Patrick's head was smaller than that of our Control's, his brain probably was no smaller than that of any human of his height. In addition, the combination of his short neck, small facial area, and carrying his cranium low and to the back of his head may have had the visual effect of him appearing to have an extraordinarily small head.

Whatever its size, this head was going to lead Patrick into trouble for the rest of his life.

*

Unusual features that Patrick may have inherited through his paternity include:

1) A radically short neck
2) Narrow shoulders
3) A likelihood that Patrick's torso was relatively long compared to his leg length
4) An elongated skull and narrow head—particularly in terms of projected frontal area and caudal or lower head area versus the "normal" Control
5) The top of Patrick's ears were *below* a plane through the center of his pupils (low-set ears)
6) A steeply sloping forehead
7) Lack of a chin—another trait of all early *Homo* species, including Neanderthal
8) By expanding his brain lower in the back and forward, Patrick's cranial physiology compensated for his relatively narrow head to reach parity or above in terms of brain volume

Patrick's measurements, derived from his mugshots, versus those of a scaled contemporary Colville man's head are consolidated in Table 2.

Metric	Patrick at 5'-2" (1.57 m) Units					Control Scaled to 5'-2" Units					Patrick vs. Control as Percent
	Inch	Cm	In2	Cm2	Angle	Inch	Cm	In2	Cm2	Angle	
Head width	6.9	17.4				6.8	17.2				1%
Head height	7.7	19.6				8.1	20.5				-5%
Ratio of head width to height	0.89	0.89				0.84	0.84				6%
Plane of acromion to underside of chin	0.88	2.2				2.06	5.2				**-57%**
Distance between acromions	9.1	23.2				11.7	29.7				**-22%**
Ratio of acromion plane height to width	0.10	0.10				0.18	0.18				**-45%**
Projected head area-side			40.0	258				41.8	270		-4%
Projected head area-front			33.5	216				37.2	240		**-10%**
Projected anterior head area-side			19.2	124				20.7	134		-7%
Projected posterior head area-side			20.8	134				21.1	136		-1%
Ratio posterior to anterior area-side			1.08	1.08				1.02	1.02		6%
Projected cranial head area-side			15.7	101				14.6	94		8%
Projected caudal head area-side			24.3	157				27.2	175		**-11%**
Ratio cranial to total area-side			0.39	0.39				0.35	0.35		**12%**
Chin to center of pupils	4.4	11.1				4.5	11.3				-2%
Ratio of pupil height to head height	0.57	0.57				0.55	0.55				3%
Jaw underside to top of ears	4.0	10.2				4.5	11.4				**-10%**
Ratio of top of ear height to head height	0.5	0.5				0.6	0.6				-6%
Head long axis	8.3	21.1				9.0	22.9				-8%
Head width at circumference	6.4	16.2				6.4	16.3				-1%
Ear lobe to front of lips	3.6	9.1				4.5	11.5				**-21%**
Angle from vertical: forehead slope					38					26	**46%**
Angle from vertical: brow to extreme back					99					91	9%
Angle from vertical: pupil through ear					121					105	**15%**
Angle from vertical: long axis of ear					41					22	**86%**
Angle from vertical: rear of cranium base					46					24	**92%**
Projected brain area: side			14.97	97				14.99	97		-0.1%
Projected brain area: front			15.53	100				14.98	97		3.7%

Table 2: Patrick Features vs. a Control

CHAPTER SEVENTEEN

Sasquatch as Indigenous People—Culture

> "There are several lines of evidence that inform us about hominid diets, lifestyles and health, including studies of living primates, archaeology, paleontology, trace element analysis and living hunter and gatherer societies."
>
> —Bogin and Rios, 2003.
> *Rapid morphological change in living humans: implications for modern human origins*[381]

The study of Patrick's features suggests that his mother's abductor looked much different than us. Even so, a Skanicum was able to interbreed with his modern human mother.

In Chapter 9, *The Type Specimen*, I repeated the definition of *hominin* as a member of either the genus *Homo* or of *Australopithecus*. I believe that the genetic introgression that Patrick and his children represent satisfy the requirements for Sasquatch inclusion into *Homo*. Skanicum is our kin, perhaps our literal "elder brother" as some North American tribes refer to

[381] Bogin and Rios 2003, page 80

them.

Sometime in the last few years I heard or read the suggestion that a sensible and humane way to approach the Sasquatch mystery is that of approaching an uncontacted tribe. Many go a step further and suggest that Sasquatch are uncontacted and also want to stay that way. Mostly, they do not want to be our friends. Noël agrees: "These are a people who do *not* want to be 'found.'"[382] Fortunately, they, *mostly*, have a live and let live approach to their dealings with their more impulsive, fearful, and violent younger brothers.

Forth writes that there are estimated to be more than one hundred uncontacted tribes of *H. sapiens* on Earth.[383] There could as well be similar numbers of uncontacted Sasquatch "tribes" on this Turtle Island and in Eurasia. We would be showing wisdom if we treated Sasquatch with the respect and caution that we approach any other uncontacted tribe of self-aware people.

Sydney Possuelo, through his work in the Brazilian government, made first contact with *nine* previously uncontacted tribes in the Amazon. Eventually he created a special unit in the Indigenous Affairs Department with a policy of *not* making contact with uncontacted tribes. Now he asks, "When there is contact in the future, will we be more brotherly, more human, less violent?"[384]

Reasons that an uncontacted tribe might want nothing to do with us include overwhelming negative cultural influences that may be introduced through foreign cultures; introduced disease; the eroding of cultural vital force through "self-domestication" with the ease of the new lifestyle; and concern that the contacting culture is looking for profit, entertainment, or conversion. Anthropologists now recommend a more circumspect approach that encourages protecting the lands of uncontacted tribes and studying these tribes from a distance without disturbing their daily routines or lifestyles. This new approach does not include chasing or forcing contact, but instead allowing the uncontacted to initiate

---

[382] Noël 2009, page 205

[383] Forth 2022 page 214

[384] Eede 2009: *We are One: A Celebration of Tribal Peoples*

interaction on their own schedule and on their own terms. Uncontacted tribal members that do approach "civilized" outposts usually are looking for food or medical care. Those approached are encouraged not to attempt to change or "improve" uncontacted tribal culture or lifestyle.

These slowly learned lessons may be useful to apply in our contacts with Sasquatch. Treat the Night People as an uncontacted tribe that so far has been determined to remain uncontacted. Protect their lands, give them space and privacy, and when they do approach, provide any assistance you are asked for in the manner it is requested, if you are able. Be grateful for efforts to repay the favor, even if the repayment is a deer leg on your porch.

Ultimately, the people making contact are likely to be the people that most closely approach the Sasquatch lifestyle: Those living quietly in the woods, taking care of family, and not looking for any trouble from the neighbors.

In our search for understanding of how these Night People are successfully living their lives, it is useful to look at previous and existing hunter-gatherer cultures of the Earth. One of my primary sources for Indigenous knowledge comes from Josephine Flood's 2010 edition, *Archaeology of the Dreamtime: The Story of Prehistoric Australia and its People.*

Much of her work references the late Pleistocene: From 50,000 years ago until the end of the last Ice Age, about 11,700 years ago. As a model for the cultural elements that we might expect from Sasquatch, Flood points out that late Pleistocene Aborigines (the indigenous pre-European contact peoples of the Australian continent)—including the Tasmanians on the southern island of Tasmania—left behind art painted on rock in caves, on boulders, and on rock faces.[385] Some of the stick structures attributed to Sasquatch seem to have no utilitarian purpose. Could they be a type of art?

Flood notes, "...Aboriginal men and women living a traditional life have more leisure time than is available to the average farmer or office-worker..." She continues, "The nomads scorned by the early white settlers were poor in material possessions but rich in

---

385 Flood 2010 revised edition, pages 125 and 137

spirit, leading a secure and healthy life ideally suited to their environment."[386] Instead of being constantly occupied with punching a clock to make a living, many Sasquatch groups might have time on their hands and the intellectual ability to appreciate art and its creation.

We get hints of what else to look for in what Flood reports of Aborigine manipulation of their environment:

> A wide range of types of arrangements, or alignments, of stones occur. These are circles, lines, 'corridors', single standing stones or piles of stones heaped into a cairn.[387]

Bruce Pascoe, in his 2018 paradigm-shifting book, *Dark Emu: Aboriginal Australia and the birth of agriculture*, is another fascinating source for the possibilities in peoples that use natural materials to express their culture:

> Villages and cemeteries were often marked with carved and painted tree trunks and timbers... A variety of structures were erected over graves... arbours that we can assume had been planted to enhance the aesthetic. ...Trees within the grove had been altered by lacing one limb over another while the trees were still saplings, so that as they grew the limbs fused and left oval-shaped windows or rings.[388]

The passage above gives those of us poking about in the woods many things to look for as evidence of the presence—past or now—of Sasquatch, such as marked tree trunks and intertwined trunks and branches.

Flood's words suggest that we should not discount the possibility of trade in goods and knowledge between different areas of Forest People:

> Aboriginal tribes were not isolated groups, but part of a

[386] Ibid., page 266 and 267
[387] Ibid., page 275
[388] Pascoe 2018, pages 137-138

> complex social and economic network. Nearly all communities traded with their neighbors, and this exchange system served to pass on not only goods, but also ideas.... Change could come about through the diffusion of an idea as well as of the actual artefact.[389]

Flood continues on to suggest that, "'exchange' and 'distribution' are better terms" than the word "trade" for activities whose primary importance may be cementing social ties between neighboring groups that are, "based on social and ritual needs as well as utilitarian requirements... exchange might well be rooted in systems of reciprocal gift-giving, rather than a need for raw material or a desire for exotic goods not available locally."

In other words, the trade that Sasquatch might be engaged in between themselves, *or with us*, may look more like gift-giving based on qualitative significance rather than exchanges of perceived equal value. And that is *exactly* what we find in the "gift giving" phenomena so many people report experiencing with the Night People. A blue feather or a clean deer skull might be left in exchange for a borrowed tool.

Marlowe writes, "...it doesn't seem particularly unusual that a primate as sophisticated as Bigfoot could conduct rudimentary communication or trade with human counterparts."[390]

In my "trading," I have left dog treats, organic apples, chocolate, peanuts, and toys. In the last year I have gotten back a broken rock, an old campfire grill, a white pebble, and a battered water bottle. It's literally not the value of these trades or gifts—*it is the thought that counts*! They are signaling their desire for a peaceful co-existence with us, their semi-hysterical neighbors, by leaving these sometimes confusing tokens. A torn animal leg on your porch is not a threat. Think of that deer leg as a plate of chocolate chip cookies fresh from the oven.

The fact that we are getting trade gifts from Sasquatch now is enormously promising. We have an opportunity for peace between our tribes if we have the courage to accept our wild

---

389 Flood 2010 revised edition, page 268

390 Marlowe 2013 page 10

neighbors for what they are and for what they offer to us.

It has been noted that when we leave foodstuffs at gifting stations, sometimes only half of the item or items are taken—a possible recognition of the value of what has been offered along with a polite restraint on the part of our shy trading partner. For those experiencing some success in "gifting," it may be helpful to attempt the same approach back. For instance, if finding deer remains on a porch, try a slow outdoor-barbeque of the meat (bound to attract attention if you have Forest People around). Show trust by eating some of the meat and leave half the cookout 100' into the woods. World Bigfoot Radio's *Glagg Saga* implies that cooked meat may be greatly appreciated.[391]

Even on a continent smaller than North America (Australia), Flood sees language quickly evolving apart in widely separated groups: "...languages change at such a rate that after 3000 to 4000 years of separation, genetic links [in this usage, linguistically conveyed connections] are no longer visible."[392] Grossinger did not see language drift as a barrier for Nahganne to communicate to distant populations of their own: "...if sasquatch understands some simple things, such as hand gestures, they must then have an ability to communicate with each other and share ideas, thoughts, and other traits related to intelligence…"[393]

The capacity for language, though immediately an acknowledged ability among the Pleistocene Aborigines, is still a debated trait of Sasquatch, but a great deal of evidence exists that these Original People do employ complex language(s), using a variety of sounds to convey information.

The sounds of people in the woods talking too quietly or indistinctly to be heard ("mumbling") are often reported. Grossinger described whisperings[394] and Isdahl a "...melodic whistle kind of chatter..."[395] Short corroborates with a report that

---

[391] World Bigfoot Radio podcast; *Glagg Saga Complete*; https://www.youtube.com/playlist?list=PL2islruoPuKrB4xiXj1Op8I5GOqplngj1

[392] Flood 2010 revised edition, page 235

[393] Grossinger 2022, page 154

[394] Ibid., page 18

[395] Isdahl 2021, page 53

the northern Hare branch of the Athabascans in northern Canada and the Gwich'in of the Bonnet Plume River of Yukon both believed Bushmen communicate with whistles.[396]

Whispering and whistles are just some of the language-like sounds reported. Beginning in the early `70s, Ron Morehead, Al Berry, and a few more hunting partners recorded sounds, some apparently language, some even dubbed "samurai chatter." These recordings are considered the most diverse and best ever made. Clips may be located on the Internet under the name, "Sierra Sounds" or at Ron Morehead's site.[397] The sounds have been described as being made by a vocal tract that can reach both higher and lower pitches than modern humans. Crypto-linguist, R. Scott Nelson, concluded from a study of the tapes that these sounds were not created by *H. sapiens*, that the sounds are of a complex language, and that the tapes could not have been faked.[398]

Witnesses tent camping in the southern Washington Cascades, after experiencing various indignities I already wrote about, heard "the craziest language."[399] The same author (Isdahl) reports later in his book that boys camping in Tennessee were to hear "ungodly chatter" and screaming.[400] Not all the accounts that Isdahl repeats are so unsettling. In one account a witness told of a male Sasquatch appearing to talk into the ear of a female.[401] Alley repeats an Alaskan report of sounds like "two old Finlanders chattering away."[402]

Native American tradition that goes beyond whistles as communication include the Chickasaw of Mississippi, Alabama, Tennessee, Oklahoma, and Kentucky. Short described the 'Lofa" as a smelly, hairy being that spoke a language.[403] The Chehalis tribe of British Columbia, as told to pioneering researcher John W. Burns, accepted that the Sasquatch speak their own Douglas

---

[396] Short 2024, pages 105-107
[397] https://ronmorehead.com/ [accessed 2026.01.30]
[398] Morehead 2017 (second edition), pages 143-150
[399] Isdahl 2021, page 150
[400] Ibid., page 269
[401] Ibid., page 65
[402] Alley 2003 (2018 edition), page 194
[403] Short 2024 pages 107-108

language or a similar dialect.[404] Scott Marlowe acknowledged in his book that the Albert Ostman account implied the Sasquatch family were able to communicate with each other.[405]

A shared language, at least locally, may also include names for each other according to Frank and Noël.[406]

Observed Sasquatch culture, similar to reports of human hunter-gatherers, goes beyond language. Scott reported of the activity around her home in Washington that there was a male in charge of this group and that he was determined that all important communication go through him.[407] Noël confirmed this patriarchal group structure and wrote, "In Sasquatch groups, an alpha male tends to be in charge." and that these male "protectors" will deal with armed humans as necessary.[408] One Grossinger witness described a pregnant female Sasquatch in early July 2008, maneuvering herself behind her male companion, apparently for the protection she believed that he provided.[409]

In another instance of the strong protecting the less strong, in the same Grossinger book, two witnesses describe driving south from Yukon towards Haines, Alaska when the car they were in slid off the road in a snowstorm and became stuck. After they had been working for some time to get unstuck, they noticed a seven-foot-tall creature "standing and watching" them from the top of a close low cliff. One of the men retrieved a machete from the vehicle and brandish it threateningly at the watcher. The seven-footer left. However, it reappeared five minutes later, but now standing behind an even more imposing nine-foot Sasquatch that made several threatening moves towards them from the top of the cliff. The men quickly got back into the stuck car and locked the doors.[410]

It seems that Skanicum tends to live together in groups, especially for child rearing. Flood reported that pre-European

---

[404] Ibid., page 82
[405] Marlowe 2013 page 30
[406] Frank 2021, pages 57 & 98; Noël 2019, pages 12 & 117
[407] Scott 2010, page 57
[408] Noël 2019, pages 17 & 39
[409] Grossinger 2022, page 154
[410] Ibid., pages 161-162

contact Aborigines also existed in "...small, highly mobile groups…"[411]

Noël interviewed a witness who remembered a childhood encounter with an adult male and female, a teenaged boy, an adolescent girl, and a little boy Sasquatch.[412] Grossinger had a witness report, from British Columbia's isolated Cassiar Highway, of an encounter on July 3, 2000. While driving south, in the middle of the night, the witness had a roadside crossing in front of him of an adult male Sasquatch who had his arm draped over the shoulders of a female who was herself cradling an infant in her arms. They walked slowly and unconcernedly into the forest.

Things may have changed since the thorough researcher, John Green, was accumulating and studying Sasquatch encounters, but George Gill highlighted in 1980 that, "according to Green there are clearly more females and juveniles reported from the more northerly parts of the range, particularly from Canada, than from California or Oregon."[413] I have not seen other suggestions that Sasquatch may preferentially bear and raise children in the northern parts of North America. Perhaps the trend observed by Green reflected a reporting gap or greater caution on the part of southern Sasquatch populations, rather than on a lack of family structure in the south.

Scott referred to a persistent theme of occasional "rogue males" in Sasquatch society. Rogue males supposedly are younger Sasquatch or, more rarely, very old males, that live isolated from their own kind.[414] Wiitala was aware of a white Sasquatch in his area, on its own and perhaps an "elder."[415] Conjecture on my part would suggest that the rogue male phenomenon could be accidental or temporary separations from a group; young males in search of a mate; males whose family group has died for some reason; delinquent males that have been outcast from Sasquatch society; or males just out adventuring and

---

[411] Flood revised edition 2010, page 103
[412] Noël 2019, page 53
[413] Gill 1980 page 270
[414] Scott 2010, page 97
[415] Wiitala 2021 page 44

seeing the country.

Patrick's father may have been a young rogue male. Abductions of the Hairless Ones are so rare, it seems likely they are not normal behavior in Sasquatch culture.

Canvassing indigenous *Homo sapiens* habits for clues into Sasquatch social behavior, of note is Flood's point that the Australian Aborigines practiced both burials and cremations at least 40,000 years ago.[416] For any researcher encountering atypical Sasquatch behavior, she reminds us that, "concern for the dead is expressed not through buildings but through complex rituals, which may go on for weeks."[417]

Noël quotes one long-term experiencer in Texas: "...the great surge of activity I was experiencing here last spring was not a siege but a celebration, connected in some way with the ancient spring ritual called Beltane..."[418]

Flood again gives us avenues of investigation when she writes of early hunter-gatherer spirituality in Australia. "...life of the spirit is all important in Aboriginal society…" She found that landmarks in the natural world are tied to their spiritual mythologies, but that these same landmarks generally contain no traces of the peoples that found them significant.

> ...it was not part of Aboriginal culture to build monuments… Aboriginal religion takes a different form… Some totemic increase sites are marked by arrangements not of stones, but of bones.[419]

According to Flood, caves were not used for shelter or homes as much as they were used for rituals.[420]

There is controversy also on the subject of Sasquatch and fire-making and use. Some Native American traditions do describe Sasquatch as using fire. Short writes, "There is much testimony from surviving concubine captives (both First Nation and Native

---

[416] Flood 2010 revised edition, pages 204 and 281

[417] Ibid., page 273

[418] Noël 2009, pages 196-197

[419] Flood 2010 revised edition, 268 and 273

[420] Ibid., page 281

American women) that the Sasquatch indeed uses fire to light deep underground caverns and cook by."[421] However, some Native American traditions of Sasquatch habits *do not* include fire use. In all Alley's many years of thorough and thoughtful research, he found no evidence of fire use in Sasquatch.[422] Perhaps fire use was a skill of North American relict hominins in the past, for its convenience and comfort, but that skill has now been largely lost or abandoned due to the necessity of concealing family units and core territory from the Hairless Ones.

Another hunter-gatherer observation challenging to our preconceptions is that of interspecies cooperation. Sasquatch is sometimes suspected of mutually beneficial arrangements with coyotes and ravens. Although the notion seems initially almost fantastical, the Australian Aborigines apparently developed even more startling partnerships with other species—at times, with dolphins per Pascoe:

> Foster Fyans, a police magistrate, saw Aboriginals fishing at Geelong in partnership with dolphins who drove the fish in to the shore. Similar relationships were reported at Moreton Bay and many other Australian beaches."[423]

Flood devoted considerable attention in her book to how the isolated Tasmanians dealt with a cooler climate on the southern island of Tasmania. Initially, the Tasmanians adapted to the cold by sheltering in deep caves, eating a high-protein diet, using fires, wearing simple clothing, and consuming lots of calories. As the climate moderated, the Tasmanians made the unusual switch to abandoning clothes. When the temperature still occasionally dipped uncomfortably low, they smeared animal fat and ochre on their bodies as a minimal coating to protect them from the cold.[424] This accommodation caught my eye, because there are many reports of persistently greasy finger, hand, and face prints being left behind when Sasquatch touch autos, auto glass, and house

[421] Short 2024, page 96
[422] Alley 2003 (2018 edition), page 277
[423] Pascoe 2018, page 71
[424] Flood revised edition 2010, page 138

windows. Perhaps they are simply smearing animal fat on themselves just like the Tasmanians for warmth, or even to keep their skin from cracking.

Based on the Tasmanians, Flood suggests that the lower limit for habitually unclothed humans (of our variety anyhow) is about -05 degrees C (23 F).[425]

Tom Higham in his fascinating 2021 book, *The World Before Us: The New Science Behind Our Human Origins*, defines the type of places we may expect to find hunter-gatherers—as have others—simply through annual rainfall patterns: "Hunter-gatherer groups cannot cope in environments where rainfall is less than 90mm [3-1/2"] per year; it's just not possible for them to survive."[426] Flood adds her perspective from the arid continent of Australia when she writes (quoting M.A. Smith et al from 1997) that one necessarily finds a "pattern of land use tethered to reliable water sources."[427]

In considering our relationship to Sasquatch as that of an uncontacted tribe in our midst, we must approach them as equals with respect and dignity: A proud and independent people that have chosen a different cultural path, away from our highly consumptive and ultimately unsustainable reliance on technology to overcome and master environment and ecological systems. Pascoe warns us of a potential pitfall in our perception of Sasquatch as a repeat of the European's initial attitude towards Australia's first *H. sapiens* inhabitants: "...the image of Aboriginals as primitives haplessly wandering across the face of the earth allowed Australians to feel 'sorry' for Aboriginals and dismiss them from the national consciousness."[428]

Instead, as knowledge has deepened of their very complex cultural systems, it has become clear that the Aboriginals practiced a sophisticated stewardship of the lands they lived upon, with clear areas of responsibility for every individual, family, clan, and tribe. Sasquatch may be engaged in a similar intimate relationship with the land to this very day. Our ignorance of their

---

[425] Ibid., page 138
[426] Higham 2021, page 22
[427] Flood revised edition 2010, page 103
[428] Pascoe 2018, page 155

traditional stewardship responsibilities with place may explain many of the conflicts and frustrations that the Night People have with us:

> Aboriginal law insisted that the land was held in common and that the people were the mere temporal custodians. Individuals were responsible for particular trees, rivers, lakes, and stretches of land, but only so these could be delivered forward to the next generation in accordance with law. Individuals and families might be said to own a particular fish trap or crop, but they worked it in cooperation with the surrounding clans.[429]

The impact of the relentless European invasion of North America on Sasquatch populations appears to be virtually identical to the tragic history of European impact on Aboriginal culture, as expressed in the diary kept by Sir Thomas Livingstone Mitchel on his 1831 expedition north from Sydney, Australia:

> These unfortunate creatures could no longer enjoy their solitary freedom; for the dominion of the white man surrounded them. His sheep and cattle filled the green pastures... the stranger came from distant lands and claimed the soil. Thus these first inhabitants, hemmed in by the power of the white population, and deprived of the liberty which they formerly enjoyed of wandering at will through their native wilds, were compelled to seek a precarious shelter amidst the close thickets and rocky fastnesses which afforded them a temporary home, but scarcely a subsistence...[430]

---

[429] Ibid., page 199

[430] *Expedition of Sir Thomas Livingstone Mitchell 1831*; https://freesettlerorfelon.com/sir_thomas_mitchell_expedition.htm [accessed 2025.11.18]

CHAPTER EIGHTEEN

Sasquatch as Indigenous People—Food

I reviewed food sources briefly in the chapter, *What is Sasquatch*? But we may get new ideas of what to look for in Sasquatch habits and habitats by looking at *H. sapiens* hunter-gatherer approaches to getting calories. I will look first at hunter-gatherer food sources, then compare and contrast those with what witnesses are saying about Sasquatch-type beings.

Deliberate ecological resource harvest planning and husbanding by the Australian Aboriginals was noted by Pascoe, along with purported techniques that essentially were terra-forming vast stretches of the Australian continent when Europeans made landfall. Despite the evidence of intelligent design in front of them, the white explorers self-servingly declared the continent *terra nullius* (land belonging to no one).[431]

One of the telltale signs of previous Aboriginal occupation of a piece of land, as noted by Flood, are unusual concentrations of food-bearing plants. Seeds tended to be dropped at campsites.[432] Pascoe notes the same giveaway of indigenous habitations:

> Harvesting on a cyclical mosaic over two to three years

[431] Pascoe 2018

[432] Flood 2010 revised edition, page 260

> would see no diminution of supply, but would instead fertilize and enhance the crop. These management practices created anomalous vegetation distributions.[433]

Similar "anomalous vegetation distributions" may be an indication of past or ongoing Night People use of land that we aren't now aware enough to notice.

Based on Aboriginal habits, another sign of Sasquatch activity may be harvested plants left to dry in the sun. Per Pascoe: "When in 1845 Stuart first saw the harvested grass... he found it spread out to dry and ripen on the sloping banks of a stream."[434] Grasses, although occasionally reported as being consumed by Sasquatch, may be an underappreciated potential food source for the large-gutted Skanicum: "Researchers think there are over 140 Australian grasses that were harvested by Aboriginal people."[435]

Introduced livestock in North America may have impacted Sasquatch's sources of vegetation just as it did in Australia. According to Pascoe:

> The loss of this dietary component [an Australian plant resembling clover] was crippling to the Aboriginal economy... introduced stock zeroed in on these plants wherever they grew, and as a consequence the Indigenous people lost both their habitation sites and one of their principle sources of sustenance.[436]

Short cited one report of consumed corn *stalks* in association with a Sasquatch sighting.[437] Such a dietary choice immediately puts a separation between Sasquatch and *H. sapiens*: Corn stalks are virtually indigestible for modern humans. I do recall my scat find that was filled with dense, woody fibers and also Meldrum's suggestion that the huge torsos of Sasquatch might conceal a gut capable of digesting cellulose. Noël suspects as much:

---

433 Pascoe 2018, page 24
434 Ibid., page 31
435 Ibid., page 64
436 Ibid., page 35
437 Short 2024 pages 217-218

"...Sasquatch probably eat wood; the bark in particular is rich in vitamins, minerals, and medicinal qualities."[438] Alley describes a witness seeing a Sasquatch pulling bark off a hemlock (purpose unknown).[439] Many authors, including Paulides[440], report scat containing vegetation that would be challenging or impossible for "civilized" humans to digest. If a fiber-rendering gut is the case for Sasquatch, winter food sources become less problematic when the twigs and bark of deciduous trees and shrubs and the needles of evergreens are all potential food sources. Alley reported scat much like a bear's in appearance and in huge volumes matching the largest salmon-eating coastal brown bear.[441]

According to Albert Ostman, he saw only vegetation consumed by his four hairy captors in 1924:

> They might eat meat, but I never saw them eat meat, or do any cooking....and the plants with sweet roots on the mountainside might have been in season this time of the year. They seem to be most interested in them. The roots have a very sweet and satisfying taste.[442]

Fruit consumption and theft from orchards by Sasquatch are often reported. Overnight in the late summer of 2024 on a farm I worked at in western Massachusetts, two peach trees and two plums had a couple hundred pounds of fruit spirited-off site with no broken branches or pits left on the ground. Even already fallen and partially-rotted fruit was gone. There were two personal twists to this theft: One of the plum trees was less than twenty feet from my back door, yet I slept soundly that night. And there was a vertical sliver of peaches left behind on one of the trees *precisely* where I would daily remove past-prime fruit to feed the chickens.

Pascoe tells us about an unusual food resource for the Aboriginal Australians:

---

438 Noël 2016, page 133

439 Alley 2003 (2018 edition), page 43

440 Paulides 2017, pages 147-148, 246-245, 362

441 Alley 2003 (2018 edition), page 2976

442 BFRO Report #1091 2000

> The great gatherings of people in the Australian Alps were made possible by the summer arrival of massive numbers of Bogong moths. ...Crows assembled to take part in the feast, and they too became fat and so intent on their hunt that the tribes knocked them on the head and ate them, a great delicacy, as the meat was plump and aromatic after the bird's diet of moth fat. ...The richness of the diet is attested to by settlers who witnessed Aboriginals returning from the moth harvest in great health, their bodies glistening with moth fat. ...50 to 60 per cent of the moth's weight is fat.[443]

It is known that grizzly bears will climb into the high country of the Rocky Mountains of North America in early Fall to feed off moths rich in protein and fat. There is reason to believe that Sasquatch would also harvest this valuable food source. Note the mention in the excerpt above of human bodies glistening with fat. The occasionally reported greasy finger and face prints left behind by Sasquatch may just be an incidental effect of such successful feeding.

Lapseritis has reported such varied foods consumed by Sasquatch as including, "bread, trout, apples, grain, cucumbers, cabbage, mushrooms, deer, elk, used salt blocks" and even Hershey bars with almonds.[444]

From archaeological finds, Flood describes consumption of shellfish from both fresh and saltwater sources. Shellfish consumption is revealed with the presence of midden piles.[445] Pascoe points out that among traditional users, these piles may have great antiquity:

> Western Victorian Aboriginal communities had been trying for thirty years to have an ancient midden on the Hopkins River analysed. It is so old, it has turned to rock. When that examination was finally undertaken, it was found that the midden was as old as 80,000 years, 10,000 years before the Out

---

[443] Pascoe 2018, pages 150-152
[444] Lasperitis 1998, page 197
[445] Flood 2010 revised edition, pages 113 & 130

of Africa theory says humans began to leave Africa.[446]

I recently heard a podcast detailing an event among a nighttime Canadian First Nations' harvest. They were rousted from a clam bed by an approaching ruckus in the woods above the beach. The diggers fled the beach by boat, believing they had upset the local Hairy Ones. Even individual clam beds may have been in centuries-long use by local Sasquatch—a use they may be prepared to defend as an established "senior right."

I was long puzzled by a scattering of bleached-white freshwater snail shells on the shore and in the shallows of a large lake in Denali National Park. Only recently have I realized that though this may simply be evidence of feeding by river otters, the empty shells may also be a sign of the Boreal Bigfoot.

Sasquatch is occasionally reported swimming vigorously and effectively in both fresh and salt water, sometimes for great distances to access islands. Guermer, of the *Sasquatch Chronicles* podcast, tells of an older woman reporting that she sometimes sees "monkey men" swimming the very wide Columbia river. Short reports on a man seeing three Sasquatch in 1990 emerging from cold Puget Sound onto the shore of Portage Island, near Bellingham.[447] Alley had a detailed description of one Sasquatch swimming:

> They could see it under water swimming like a frog, arms forward over its head but not doing a crawl stroke. The legs kicked, the best description was like a frog.

Any proficiency in water is almost certainly gained through the search for food, a skill quite likely practiced by both genders:

> Diving for crayfish [in this southern hemisphere, "crayfish" is a saltwater crustacean related to lobster] and abalone was an important part of the southern coastal economy. Not all the diving was done by women, but in Tasmania and Victoria the

---

[446] Pascoe 2018, page 41

[447] Short 2024, page 135

> shellfish was collected predominantly by women... The shellfish [abalone] was a favoured-high protein item of the coastal diet in this part of Australia.[448]

Flood also mentions boulder and rock constructed fish-trap pools both on beaches and in rivers designed to concentrate and trap fish as waters recede. She wrote that fish traps were in use as early as at least 30,000 years ago in Australia.[449] Dating from this same era, Flood describes the consumption of fish, shellfish, small mammals, reptiles, birds, and emu eggs. Food-gathering typically involved a division of labor along gender lines with women

A potential river fish trap found by the author in western Massachusetts in 2024. Jacket added for scale.

[448] Pascoe 2018, page 91

[449] Flood 2010 revised edition, pages 54, 201, & 232 (photo)

gathering plants, shellfish, and small animals, and men fishing and hunting the larger game.[450]

Guermer, on *Sasquatch Chronicles*, has noted repeatedly that fishermen piss-off the forest locals more than do game hunters. It seems that the Sasquatch consider the fish in the waters more *theirs* than they do the free-ranging mammals or birds. Whether it is a matter of ownership or dietary importance, I don't know.

Perhaps both.

Fish are likely as important to Sasquatch as they were to the Australian hunter-gatherers. Per Pascoe: "Permanent fisheries were an established part of the Aboriginal economy from one side of the country to the other." The notorious Sasquatch objection to people fishing in lakes and streams may reveal more about *their* culture than about ours. Fishing locations may have an "ownership" of sorts established hundreds of years ago, or more. When we violate these clan rights, we disturb the social order established precisely to prevent conflict.

In 2019, at the International Bigfoot Conference in Kennewick, Washington, one of the speakers (my apologies—I have forgotten his name) told a story of returning to a frozen stream in eastern Washington state (Patrick's country) that he used to fish when younger. Near the stream, he found a rudimentary debris hut. Inside were fish so fresh, some of them were still flopping about in the cold air. I was astounded by this story and approached that speaker later, asking for more details. He told me that when he used to fish there in his younger days, he often felt a presence watching him. He got in the habit of leaving a couple fish behind on a rock and was never bothered by the watching presence. He acknowledged that when he found these fish in the stick structure that he felt that they were a return gift for him and that he still regretted not acknowledging the gift and taking the fish with him. But the story gets stranger still. Not only is it inexplicable that some of the fish were still alive, but the river was *frozen* over! He had no explanation for the astounding circumstances and I told him that I could see why he had skipped over this detail when he told the story on stage.

[450] Ibid., pages 259 & 261

Short repeated a belief that Sasquatch travel from Canada down to Palmer Lake in Washington state to dive for bass[451] and Wiitala believe he knows of a spot in southwest Washington where several families of the Night People consistently gather to harvest from a salmon run.[452] Grossinger reported a witness seeing a Sasquatch sitting or squatting in a creek, while it caught two salmon. It killed kill each fish before throwing them onto the shore.[453]

Pascoe is a rich source for types of food stored and the preservation techniques in the Land Down Under:

> Innumerable commentators came across the preservation of everything from fish, game, plums, caterpillars, moths, quandong, figs, seeds, and nuts, among a wide variety of other foods. Preserved caterpillars were made into a kind of flour; figs and quandong were pulped and mixed to form a product that can only be likened to quince paste.[454]

Pascoe has yet more on the great variety of foods eaten and stored for lean times by hunter-gatherers:

> Stores of fish meal and fish flour were recorded, but many other commodities had their individual preparations prior to storage, including caterpillars, witchetty grubs, grasshoppers, meat, and liver. Such stores were often covered with the ashes of particular woods...[455]

Flood considered the caloric strategies the Tasmanians employed to cope with a cool climate. "Hunters in high, cold latitudes need foods rich in fat, yielding high energy. Thus, for the Tasmanians, seals and sea birds were better than fish or shellfish."[456]

---

[451] Short 2024, page 94
[452] Wiitala 2021 page 32
[453] Grossinger 2022, page 98
[454] Pascoe 2018, page 55
[455] Ibid., page 149
[456] Flood 2010 revised edition, page 206

Flood pointed out that "the most stressful time of year [is] from late winter to early spring."[457] Meldrum introduced the possibility that humans (and by extension Sasquatch) may have employed a strategy of minimal activity to cope with winter cold and reduced food sources:

> ...indigenous human populations of the past reduced physical activity during lean periods and relied upon stored foodstuffs to see them through the winter months... The possibility of hibernation or periods of inactivity is one that may have been realized in sasquatch evolution. It has even been suggested that humans show residual effects of a past history of hibernation in the form of seasonal affective disorder (SAD)... Whether hibernation is ancestral or newly developed, the genes required for hibernation are widespread among mammals. Recently, geneticists have announced the discovery of these genes in the human genome.[458]

Meldrum quotes a Hoopa medicine woman, near the west coast of northern California, as saying that many traditional food sources actually become *more* abundant in their winter woods.[459]

In Burtsev's impressive, if eclectic volume from 2015, he quotes Dr. John Bindernagel of Canada from the paper Bindernagel presented at the October 5-8, 2011 International Scientific-practical Conference on Hominology in Moscow and Kuzbass, Russia—*The Ecology of an Uncatalogued Hominoid of the Boreal Forest (Taiga) of North America and Eurasia*: "Since, even in Summer, this biome appears to provide only meager food resources, how then does this hominoid manage to survive the winters in a region in which winter conditions are characterized by severe cold combined with significant snow cover?" Bindernagel answers his own question with a discussion of different food-gathering skills, including—foraging; predation; food stealing; overwintering strategies (hibernation or torpor); storing food in caches; fat storage within the body; and migration.

---

[457] Ibid., page 134
[458] Meldrum 2006, page 189
[459] Ibid., page 191

Bindernagel continues in the same paper to mention reports of Sasquatch foods including ground squirrels, mice, ducks, deer, reindeer (in Russia), salmon, and cockles [small clams].[460]

Short is another source of the winter food storage and torpor evidence when she repeats that the *Black Giants* appearing only rarely in winter is explained by the Gwich'in First Nations people of the Bonnet Plume River of Yukon Territory as owing to its practice of hunkering down with its stored foods.[461]

Whether it is through winter migration, sheltered torpor, cached food, or something else, there is clearly a gap between the level of reported summer activity and evidence found during the winter. Meldrum points out that according to John Green's research, Sasquatch is "not active in cold weather. Less than ten percent of reports mention snow, and tracks in snow are rare."[462]

Ironically, practically the only big tracks I have ever found were in half an inch of snow on ice in *January* in Interior Alaska—a highly improbable time and place to find tracks of a shoeless naked-hominin. [Incidentally, the tracks were only five or six miles from the ill-fated Chris McCandless' bus refuge.] However, the tracks dated to a recent rare mid-winter chinook thaw, with temperatures approaching 40° Fahrenheit (4 C). And there were large numbers of caribou in that area.

The extremely adaptable Sasquatch may use a variety of winter strategies to maximize its success just as any hunter-gathering indigenous people would. In some warm climates they may remain active, but alter their food sources as needed, perhaps switching to higher calorie foods or to tubers as green plants die back. In temperate climates they may move to lower elevations and even live on the outskirts of our home-dwelling populations to scavenge much like foxes, coyotes, and ravens. Or others in the temperate zones may move closer to beaches in cold weather to take advantage of winter low tides and the intertidal zone, rich in shellfish, seaweed, and crustaceans. In colder areas still, they may migrate out of the area or indeed store food for the winter, make

---

[460] Burtsev 2015 pages 133-137
[461] Short 2024 pages 106-107
[462] Meldrum 2006 page 212

greater use of caves, and even go into a state of fat-conserving torpor. Noël describes finding a cave in a wooded high-altitude region of Arizona that had an entrance tunnel carefully cleared of rockfall debris, terminating in an inner bed-chamber with a thick layer of grass and leaves serving as a mattress.[463]

Fish is likely as important to Sasquatch as it was to the pre-contact Australian Aboriginals, but red meat and organs are also sought after. According to Higham, plant foods are a significant part of the diet for modern humans in warmer climates, but meat becomes more important in colder environments. The division of labor along gender lines becomes more defined with greater meat reliance.[464] Grossinger estimated that in his Far North research area, Nahganne diet is about 55% meat and 45% vegetation.[465]

Bryant & Trevor-Deutsch were in a position to examine five sets of collected scat (coprolites), three of which they were unable to identify as belonging to recognized mammals:

> The three remaining coprolite samples we have examined were collected on Mound Key in Florida. Everything about these specimens was radically different from the two coprolite specimens we examined from the Pacific Northwest. All three samples consisted primarily of remains of non-vegetal diets, only one containing a significant amount of any vegetal matter (grass stems and leaves). None of these samples contained parasites. The Florida samples were produced by an organism or organisms which ate mainly small mammals, insects, birds, and crustaceans.[466]

These fecal sample results, if from Sasquatch or the regional related (identical?) "Skunk Ape," suggest that diet may be extremely varied and under the influence of habit or opportunity. Flexibility and ingenuity allow Sasquatch to inhabit multiple ecozones, adapting to local food sources and climate.

As a sidenote, Baxter suggests that any such evidence sent to a

---

463 Noël 2009, page 45
464 Higham 2021, page 32
465 Grossinger 2011, page 166
466 Bryant & Trevor-Deutsch 1980, page 298

lab for analysis should go by certified mail, but that it is very important that you never ever "send surprise poop in the mail."[467]

In the central Colorado mountains I found an area in open pine forest where granite rocks had been excavated to form pits with a surrounding wall of discarded rubble. This rubble included some very large slabs weighing several hundred pounds. At the time, I thought these pits must have been the remains of early mining efforts, though I could find no evidence of tool use in moving the rocks or markings on the rocks themselves. I scratched my head over the size of some of the slabs moved. Now I wonder if this was simply evidence of Hairy Ones uncovering rodents.

According to Forth, "Ape Men" on the Indonesian island of Flores are blamed by the locals for the domestic pigs that are killed.[468] Forth believes that these reportedly small and hairy bipeds may actually be *Homo floresiensis*, otherwise known in non-anthropologist circles as Hobbit Man. *Homo floresiensis* supposedly died out some 50,000 years ago.

Some Australian Aboriginal populations got up to 90% of their meat from just one species and those meat sources tended to be carefully and fully utilized even so far as the bones smashed or split to get at the fat-rich marrow.[469] Flood also found signs of careful preservation of a food source amongst the geographically isolated Tasmanians. "Despite 20,000 years of human predation on wallabies in Tasmania, they [the wallabies—a medium-sized marsupial, similar to the larger kangaroos] survived despite changing climate, suggesting the self-limitation of human harvest."[470] Robert Kryder, of Kryder Exploration, has even suggested that Sasquatch engage in animal husbandry, at times driving deer and elk herds from range-to-range as a mobile and accessible food source.

Hunting game animals appears to be very important for Sasquatch, though they may be picky about the meat they eat, often showing a preference for some internal organs over muscle mass itself.

---

[467] Baxter 2022, page 65
[468] Forth 2022, page 113
[469] Flood 2010 revised edition, pages 125, 133, & 134
[470] Ibid., page 137

Burtsev writes that the Khanty of western Siberia describe the *Maigiki* as eating a varied diet of moss, lichens, berries, pine nuts, lemmings, mice, chipmunks, fish and bird eggs, and large mammals, much like the diet of hunter-gatherers.[471] Meldrum cites John Green's statistical analysis of Green's own reports to conclude that Sasquatch is an omnivore with almost equal association with meat-eating versus vegetable matter in sightings.[472] Bryant & Trevor-Deutsch in a paper published in 1980 in the classic collection, *Manlike Monsters on Trial: Early Records and Modern Evidence,* provide more detail about the John Green report data:

> Sasquatch is probably an omnivore. …most favored by Sasquatch are: tree roots, grass, berries, deer meat, and garbage. Other foods apparently eaten in lesser quantities include: bear, sheep, chickens, cows, horses, rodents and other small animals, grubs, clams, fish, salmon, leaves and evergreen buds, grapes, flour, eggs and bacon, milk, and doughnuts. Green pointed out that nine of the reliable eyewitness reports refer to Sasquatch eating vegetal foods while only four refer to eating of meat.[473]

This majority consumption of vegetative matter may have more to do with the amount of time and the visibility needed to harvest plant matter than a true caloric importance over meat. Many sources point to meat and internal organs being a very important focus, particularly for adult male Sasquatch. Mills, Mills, and Townsend in their 2015 mouthful of a paper, *Using Biotic Taphonomy Signature Analysis and Neoichnology Profiling to determine the identity of the carnivore taxa responsible for the deposition and mechanical mastication of three independent prey bone assemblages in the Mount St. Helen's ecosystem of the Cascade mountain range*, describe finding deer and elk kills with bones cleaned of almost all flesh

---

471 Burtsev 2015 page 122
472 Meldrum 2006 page 212
473 Bryant & Trevor-Deutsch 1980, page 295

and left behind neatly in piles.[474]

In another report, a man believed a Sasquatch took a harvested deer out of the canopy of his pick-up truck (the tee-handle latches had been closed).[475] Scavenging saves energy and avoids potential injury. Burtsev repeated stories of bear and moose being consumed by the Russian version of Sasquatch.[476] Grossinger reports a hunter seeing a northern Sasquatch taking the moose that he had just shot.[477] From the same source, we have a startling report of children on a school trip to Sheep Mountain on Lake Kluane in the Yukon, seeing a Sasquatch on the mountain above them catch and run off with a Dall sheep![478]

A few years ago, while wandering in the woods close to my rural Alaskan subdivision (mainly wooded lots of five to ten acres), I found the bone remains of two full-grown moose within 25' of each other at the top of a forested hill. I didn't think too much about it for a couple of years before it hit me: There is no "moose graveyard!" One sick moose is not going to find the remains of another moose as a setting for a peaceful death. It took a couple more years before I heard that the upper area of a nearby gravel pit (not visible from the main road) is used by the local Department of Transportation to dump road and train-killed moose and caribou out of sight of the tourists. The place where I found the two moose skeletons is only about 500' away and is the local high point of ground. I now suspect that two scavenged one-thousand-pound (450 kg) moose were carried from the gravel pit to the relative cover of this wooded hilltop for relaxed consumption—a feat even a grizzly would be very unlikely or unable to tackle.

Much has been made in recent years of Sasquatch being an ambush predator: Conserving energy and minimizing the chance of injury by using lures, camouflage, stealth, and close proximity to capture unsuspecting prey. Mills, et al believe that they have found evidence of this approach in their findings of chewed bone

[474] Mills et al 2015
[475] Isdahl 2020 pages 251-252
[476] Burtsev 2015 pages 161-163
[477] Grossinger 2022, page 243-244
[478] Ibid., page 265

concentrations in the Washington state Cascade Mountains, "Comprehensive terrain analysis and the large amount of peripheral bone evidence strongly suggest repeated ambush behavior manifestation."[479]

There are rare reports of people in the woods horrified to realize that they are in a narrowing log-lined chute—attributed to Sasquatch—that leads to an apparent "dispatch" choke point. The Australians (and Alaska Natives on graveled river plains) developed these on a grand scale per Pascoe:

> The pastoralist James Dawson, and Robinson, mention game drives or 'grand battues' where people were engaged in driving game across a 32-kilometre front to a dispatch point. Dawson reports on the cooperation of several tribes involving over 2,000 participants.[480]

We can only imagine the scale of game drive that Sasquatch would be capable of.

To those of us wedded to the grocery store, let's put this hunter-gather lifestyle into perspective:

> ...they [Australian Aboriginals] also generally ate better. …more balanced, varied and nutritious than that of many white people. …One wichetty grub yields the same amount of protein as a pork chop. ...these experts of stone age economics had a healthier, more nutritious diet than many Europeans today.[481]

Unfortunately, all of this industry and food security largely collapsed when the Europeans appeared in Australia. "All manner of foods were preserved and stored, but such preservations became more difficult when the clans were forced into constant movement by the advance of the pastoralists."[482] It is possible that the same food storage collapse happened in North America to Sasquatch.

---

479 Mills, et al 2015, page 28
480 Pascoe 2018, page 51
481 Flood 2010 revised edition, pages 266 & 282
482 Ibid., page 148

If you appear to threaten any animal's stored food, you appear to threaten their very survival. Please do not disturb any stored food that you may find in the wilds and leave such locations *immediately*.

Taking an example from the habits of *H. sapiens* hunter-gatherer societies, anyone looking for Sasquatch sign might search for evidence of harvested food sources such as: Unusual concentrations of food plants (e.g., apple trees and raspberry bushes in isolated locations in the woods); large bones smashed or split for the marrow; bones, shells, or eggshells left in neat piles; fish traps appearing like deliberately constructed rings of rock and boulders in intertidal zones or in the upper reaches of fluctuating river water levels; stashes of berries, tubers, leaves, meats, etc.; unusual or atypical concentrations of ungulates that may appear and disappear abruptly; and concentrated or piled fish remains such as fins and guts.

CHAPTER NINETEEN

Sasquatch as Indigenous People—Shelter, Range, & Technologies

"The emerging picture is of highly mobile, broad-spectrum hunter-gatherers, exploiting a wide variety of marine and terrestrial resources."

—Josephine Flood, *Archaeology of the Dreamtime: The story of prehistoric Australia and its people*; 2010 revised edition, page 38

The mechanics of Skanicum life—such as how and where they house themselves and the extent of the technologies they may employ—should be considered in our search for understanding the full Patrick story. Once again, these topics can be examined both from the angle of contemporary witness reports and also by looking at the measures that past hunter-gatherer societies have used to make the most out of their environments. Higham wonders if "we might again be underestimating the skills of ancient hominins. Perhaps there was a greater degree of technical skill earlier in the Pleistocene than we assume."[483]

---

[483] Higham 2021, page 172

Flood reports that Tasmanians went almost entirely unclothed down to about -05°C (+23°F); only covering themselves with animal fat and ochre to provide a very thin layer of protection or to rarely wrap animal skins around themselves.[484]

Sasquatch continues to surprise with their almost complete lack of any reports at all latitudes that mention clothing. Pyle notes, "Only one or two reports mention deerskin capes or other adornment."[485] Though Redfern points out that crude clothing was reported in the past and up until the 1960s and `70s.[486]
We have already discussed how the bigger bodies of Sasquatch minimize their surface area, thus minimizing heat loss, but the physics of increasing volume is not sufficient to explain their reported presence in extremely cold climates. Flood also writes that archaeological evidence suggests that the same Tasmanians did use deep caves to survive freezing temperatures. They sometimes occupied specific caves and rock shelters for tens of thousands of years.[487]

It has been common for Sasquatch researchers to point to caves as an obvious explanation for how Sasquatch copes with cold and to stay out of sight so much of the time. Meldrum contends that despite this popular explanation, there is "infrequent mention of signs of cave habitation present in accumulated sasquatch reports."[488] I recall just one case in which spelunkers reported anything hinting at Sasquatch below the surface of the Earth and another case in which shocked witnesses encountered two Sasquatch in an abandoned railroad tunnel. In another case, one of Alley's witnesses described a bed of woven branches deep in a mine shaft.[489] I worked with a man who had seen a large rectangular bed of greenery in a rock shelter (a rock overhang) at a high elevation in Denali National Park in Alaska. He felt that he had seen a bear den. I was not so certain.

Alley uncovered a report from pre-1908 that he dubbed *The*

---

[484] Flood 2010 revised edition, pages 133 & 138
[485] Pyle 1995, 2017 edition, page 91
[486] Redfern 2016, pages 54-55
[487] Flood 2010 revised edition, pages 125 & 157
[488] Meldrum 2013, page 104
[489] Alley 2003 (2018 edition), page 239

Sasquatch in the Glacier. In this case, a man fell into a crevasse on the Malaspina glacier near Yakutat, Alaska. He was able to follow ice fissures downhill in the glacier until the space widened into an ice cave or cavern with a streambed floor. In the dim blue light, he encountered a hair-covered bipedal creature, "with giant form."[490] Today we would immediately identify this ice-cave dweller as a Sasquatch.

The Ostman account from 1924 describes a ten-foot-deep by thirty-foot-wide (3x9 m) rock overhang (typically called a "rock shelter" by anthropologists) furnished with a deliberate bed of dry moss and blankets of woven cedar bark.[491] It is puzzling that after 100 years, the technological sophistication of these woven blankets are essentially unmatched in reports in the century since.

Even though there is little evidence that caves are used with consistency, many of the accounts in Chapter Seven, *Abductions*, do include caves as the location for confining the abducted.

In ancient Tasmania, windbreaks and thatched huts were made, usually by the women, as simple protection from the elements.[492] These constructions—and a reconstructed Arapahoe tribe brush shelter I saw in a museum in Longmont, Colorado—are virtually indistinguishable from domed stick structures associated with Sasquatch today. It should also be noted though that the available evidence does not confirm that these stick structures are made by Sasquatch although I am inclined to believe that some of the structures are consistent with what we know of these shy people.

Pascoe describes some of the structural forms in use by Australian Aboriginals when Europeans first started exploring the continent:

> ...an array of housing styles: rectangular structures... tent-like structures... shade houses... platforms built in trees in swampy areas... and a modification of the domed structure for the wet season... Building types varied according to the materials available... Where safe cave systems occurred, they

---

490 Ibid., pages 102-103

491 BFRO 2000. Ostman Account, Report #1091

492 Flood 2010 revised edition, pages 202-203

were also used for housing and ceremonial purposes.[493]

Pascoe, again, repeats that some houses, rather than being primitive, "were often quite elaborate, similar in size and form to the teepee of the American Indian, though made entirely of local wood and plant fibre."[494] A skeletal form of the typical North American tipi is well known in the Sasquatch researcher world. It is possible that early European explorers in North America encountered structures abandoned by the Hairy Ones that they automatically attributed to First Nations people.

Pascoe quotes explorer Thomas Mitchell: "they were covered, not as in many other parts, by sheets of bark, but with a variety of materials, such as reeds, grass, and boughs." Pascoe writes that these buildings were "substantial constructions with thatches over 30 centimetres [12"] thick."[495] and repeating what Charles Sturt saw on his expedition of 1844:

> The entrance of one was 14.5 metres wide and two metres high, the roof having been plastered with a thick coating of clay. In another area, he saw huts made with strong elliptical arches, covered with boughs, and rendered with 'a thick coating of clay so that the huts were impervious to wind and heat.'[496]

There have been many stick structures and vegetation manipulations found across the globe thought to be the work of relict hominids. These include arches, beds, the de-barking of trees, nests, debris-hut shelters, lean-tos, fences and other barricades, seeming imitations of human structures including houses and tents, free-standing jambs (sticks or logs pushed vertically into the earth), ground glyphs (small to large "symbols" made from sticks or logs), tipis, massive groundworks (*logs on the ground deliberately arranged), no trespassing signs (such as Xs or logs blocking trails and roads), "playpens," upside down

---

493 Pascoe 2018, pages 121-123
494 Ibid., page 103
495 Ibid., page 109
496 Ibid., page 101

trees, and tree and twig breaks.[497]

Noël suggests a label for a subcategory of any structure: *Side hustles*, smaller and simpler versions in close proximity to their larger examples.[498] I have seen such relatively clumsy secondary creations and attributed them to youngsters imitating their grownups.

Burtsev reports stick structures in Russia that look much like what is found in North America—even near the outskirts of Moscow itself.[499]If indeed Sasquatch are sometimes responsible for these vegetation manipulations, they have given us plenty of

Lean-to found in Colorado, USA in 2023. For scale, the human in the picture is just over two meters tall (79.5"), making this structure over three meters high (10-11').

material to chew on. According to Noël as of 2019, LeeAnn of Ontario (of the YouTube Channel Southern Ontario Sasquatch),[500] had reported finding 700-800 structures without even venturing into deep wilderness. He quotes LeAnn, "I

---

[497] Sollie 2023 & 2025

[498] Noël 2023, page 267

[499] Burtsev 2015, pages 154 & 187 and personal correspondence Feb 2026

[500] Southern Ontario Sasquatch https://www.youtube.com/c/SouthernOntarioSasquatch

couldn't believe I was finding these complicated structures in these city forests so close to civilization. I just kept finding more and more."[501]

Pictured is perhaps the most remarkable stick structure that I have found yet: A densely packed crescent-shaped half-wall cupped towards this view, with two logs across the ends of the crescent tips and wing logs flanking the "entrance." I have wondered if this enormously complex structure was an attempt to imitate a human tent.

I have had similar though less numerous findings. Some examples are pictured on these pages. Much of what I find seems to be intended specifically *for* human discovery and in a couple of cases, specifically intended *for me* (a strange conclusion, I admit). I will write more about those experiences in Volume III—a deeper examination of the peculiar phenomena that sometimes accompanies Sasquatch activity.

Perhaps contemporary stick structures are a remnant of a once active and more sophisticated Sasquatch technology, now hidden from view and in a state of regression due to the presence of

[501] Noël 2019, page 126

Europeans. The stick structures used today are commonly reduced to harmless artistic works deliberately placed where hikers and hunters may find them, or, in subtle ways, as communications to each other. Examples of subtle intraspecies communications would be X-structures as "no Trespassing" signs and small ground glyphs to indicate the presence of specific individuals or family groups.

A triple teepee found alongside a human hiking trail in the United States Rockies near Nederland, Colorado on 2023.07.06. Note the apparently freshly peeled poles in the foreground. Any stick structure found alongside a popular hiking trail must first be assumed to be a human-creation. Curiously, there was still a teepee on this spot in 2025, but it had been converted to a much less remarkable single teepee. The excess poles were nowhere to be seen.

Noël, from within his theory that Sasquatch shows many behaviors associated with autistic people, has another take on the possible motivation for making stick structures:

> Most Sasquatch structures, as we have seen, are clearly insufficient as protection from the elements. Perhaps, instead, they function as occasional, embracing havens to hole up inside, affording not so much physical as psychological shelter.[502]

A very tall single teepee found near the previously pictured triple teepee (visible back right) near Nederland, Colorado 2023.07.06. It would be possible, but difficult for a single "modern human" to tip these 20' long poles into place. On a second visit in 2025, this teepee was unchanged.

[502] Noël 2023, page 124

Pascoe describes some of the construction feats in stone of the supposed "primitive hunter gatherers" of pre-European contact Australia:

> ...villages with stone walls and roofs of thatch have been recorded all across Australia... north of Port Jackson... The walls were built from stone and clay, and the floors were covered in soft paperbark and ferns.[503]

"Stone was also used to construct hides from which to hunt animals and birds, and as protection for sacred items."[504] according to Pascoe.

We shouldn't be quick to dismiss the possibility that Sasquatch (or something similar) may express technology through stone at a level beyond the simple rock stacks or cairns that are sometimes found.

"In the south-east of the [Australian] continent, whale bone was often used, as its curve produced great strength when radiating lengths were bound in the centre and thatched."[505] Whale bone is another possible Sasquatch structural component to look for in the northern coastal reaches of North America and Eurasia.

Another example from Australia of potential trace evidence of Sasquatch activity, past and present: "Huge mounds were raised in these reed marshes near Swan Hill so that villages could be located at strategic locations within the swamp."[506] Similar mounds created by Sasquatch (or the Skunk Ape) might be found in the deep swamps of the southeastern United States, likely associated with and even expanded by midden and bone piles.

Life for the technologically pared-down Tasmanians was not as challenging as you might think. The simplicity of their technology also made them rich in time: Tasmanians are thought to have spent about a quarter of their hours sedentary in

---

503 Pascoe 2018, pages 126-127
504 Ibid., page 134
505 Ibid., page 134
506 Ibid., page 56

villages.[507] Sasquatch may have similar amounts of free time, unencumbered with *stuff.*

Theories of home range sizes for "relict hominids" come up frequently. We don't even know if the concept of a home range is valid for Sasquatch, but perhaps it is useful to think in terms of carrying capacity of different biomes. Flood found that:

> In spring and summer food was usually plentiful, but winter was a time of stress, when the river was less productive of food. Then the people dispersed in smaller groups into the back country...[508]

Contemporary accounts, particularly from residents that believe the Night People are close, often include a seasonality—Sasquatch are typically predictably present some months and gone during some others. Flood's words again, suggesting a parallel with Sasquatch society:

> Groups might stay for several months at the same camp in a rich environment, but no groups stayed in the same place all year round. When food was less abundant they would move camp more often.[509]

Flood, citing Golson, Allen, and Hope (1983) describes the preferred habitat for modern humans in the Ice Age: "...Pleistocene settlement occurred mainly in mid-Montane forests accessible to 'a vast altitudinal spread of resources extending downward into lowland valleys.'"[510] The parallel of Sasquatch sheltering on wooded slopes, but utilizing resources in lower valleys, might be expanded to include accessing the resources of high altitude mountain slopes.

Alley suggested that 100 square miles of temperate coastal forest may be capable of "easily carrying one sasquatch" jumping to as high as 500 square miles in less desirable country of the same

[507] Ibid., page 202
[508] Flood 2010 revised edition, page 260
[509] Ibid., page 260
[510] Ibid., page 37

biome.[511] However, Meldrum published an unattributed quote suggesting a Sasquatch home range of about 1,000 sq miles.[512]

The range of any hunter-gatherer group will be completely dependent on the concentration of resources in their territory and the seasonal availability of food. The area requirements for grizzly bears (*Ursus arctos*) may be similar to what we might expect for that of Sasquatch. The coastal Alaskan brown bear, reaching one thousand pounds *or more*—living in a relatively mild climate in a resource-rich environment of seasonal salmon, berries, and intertidal zones—may have a range of as little as 1 square mile. In the relatively food scarce Interior of Alaska and Canada, grizzlies may have ranges of as much as 100 square miles, with an average closer to 20.

The *potential* population density of Sasquatch may be much greater than we are prepared to accept. Alley believed that awareness of *more than* 90% of Sasquatch experiences in his area never made it past family and close friends, yet he still amassed a significant store of accounts.[513] Instead of being rare and endangered, their numbers may have been increasing for the last 100 years. In 1924, the year of the Ostman account, their numbers might have been at a nadir. Ostman was apparently kidnapped as a mate for the daughter—a choice likely driven by need owing to a lack of suitable local Sasquatch options.

In recent decades, some rural residents find, after years of bucolic living, that their idyllic retirement home is now surrounded by "something in the woods." These new neighbors are capable of becoming territorial as their home range shifts or expands, even attempting to run off the existing landowners. This competition is non-violent to humans, but may include bangs on the house, dead livestock and pets, or roars and screams from the close-by woods. Wiitala writes that "I found one family [of Sasquatch] almost living in the suburb of a small town."[514] Noël has a similar perspective:

---

[511] Alley 2003 (2018 edition), page 302

[512] Meldrum 2006 page 188

[513] Alley 2003 (2018 edition), page 317

[514] Wiitala 2021 page 36

> ...they are far more numerous, widely distributed, and, in some cases, much closer at hand than we have so far grasped. Their territory honeycombs our own.[515]

There are areas that Native Americans and Alaska Natives avoid as the established territories of Sasquatch. According to Pyle's source, Cultus lake in Washington state was shunned because, "...many Sehlatiks people there. Indians afraid of Sehlatiks people."[516] Pyle collected another account from Alaska:

> Martha Demientieff, a native Alutiiq writer and teacher whose family runs a river transport company, told me of a Yukon [Yukon river] Village deserted as recently as the summer of 1992 because of the appearance of the Wood Man, sometimes known as Neginla-eh.[517]

In Alaska, these areas that are better left for the sole use of the giant occupants are avoided as no-go zones.

Concepts of seasonal ranges, overall territory, and no-go zones probably include accommodations for secure child-rearing. Lapseritis feels that they "migrate frequently but have one or two home territories where they feel safe enough to raise young and live without intrusion."[518] But Wiitala writes, "Sasquatch are everywhere there is a patch of wilderness."[519] And he believes he has detected multiple family groups foraging separately from each other, but in the same extended territory:

> There are three families that are very close and move in unison. They have about 250 to 350 square miles of territory between the three families. They move back and forth among the valleys within that territory. Within each valley they spread the three families out to give each other a little space.[520]

---

[515] Noël 2023, page 1
[516] Pyle 1995, 2017 edition, page 210
[517] Ibid., page 183
[518] Lapseritis 1998, page 58
[519] Wiitala 2021, page 51
[520] Ibid., page 54

Flood confirms this grouping of families among *H. sapiens* hunter-gatherers as well, describing a "band" as 2-6 extended family units sharing and occupying a specific range.[521] Noël, basing this next information on a secondhand telepathically received communication, wrote that one family group had a yearly travel route with an annual meet-up place for the multiple families of an extended tribe.[522] Despite the unverifiable source, this description is not at odds with the social structure of many hunter-gather societies or even with that of modern *H. sapiens* families.

Home ranges shift among hunter-gatherers as climate and food supplies necessitate.[523] Rural residents may have Night People as neighbors for years only to have them suddenly disappear literally overnight. Noël quotes Wayne from Michigan: "I don't know where they go, but they obviously go somewhere."[524]

Seasonal movements are part of many hunter-gatherer groups, particularly in temperate, arid, or cold climates as a reaction to cyclic changes in location and quality of food and water sources. It is logical to anticipate that the same may be true of some Sasquatch populations. Flood describes seasonal movements among Australian Aborigines:

> In the heat of summer, people would have stayed close to the plentiful fresh water and shellfish of the lakes. In the cooler winter, they probably spread out away from the lakes onto the arid planes and hunted land animals…[525]

Despite the presumed mobility of a creature the size and physical capacity of Sasquatch, there is little evidence of great travel distances for specific individuals, although Melba S. Ketchum has reportedly found a DNA match between samples taken in Saskatoon, Canada and in the lower United States. However, Short points out that "no data shows a northern

---

[521] Flood 2010 revised edition, pages 217-218
[522] Noël 2019, page 54
[523] Flood 2010 revised edition, page 135
[524] Noël 2019, page 64
[525] Flood 2010 revised edition, page 38

footprint in the southern states or vice versa."[526] This minimal evidence of long-distance travel suggests the possibility that North American populations may be genetically isolated from one another. Flood examined the requirements to maintain an isolated human population over large blocks of time through the example of the Keppel Islanders off the coast of eastern Australia, near the Great Barrier Reef. Thought to average only about 85 people, this population was "more or less isolated for 5000 years" from mainland Australia. Their population density amounted to about two people per kilometer of coastline. Their material culture in that period of isolation—like that of the Tasmanians in their much larger territory—became increasingly simple.

Genetically speaking, "the limiting viable [long-term] population may be somewhere in the range of 400 to 600 depending on local circumstance and the vagaries of chance." Not coincidentally, Flood points out that 500 is also about the average "tribe" size in Australia.[527] Perhaps Sasquatch has similar regional population-size groupings with shared linguistic and cultural traits within those groups or tribes.

It is possible, or even likely, that North American Sasquatch groups went through the same drastic population declines—as did the First Nations peoples—when wave after wave of European diseases, including the particularly lethal smallpox, traveled *ahead* of the invaders all the way across the continent by the 1770s. This wave of death did not require the in-person advance of Europeans, but propagated in front of the whites like ripples from a rock dropped into a pond.

The new pestilences that the indigenous North Americans (perhaps including Skanicum) were subjected to included influenza, malaria, measles, pneumonia, scarlet fever, smallpox, typhus, and yellow fever. Death was not simply a matter of one epidemic wiping out a tribe with no established resistance (dubbed a "virgin soil epidemic"). Instead, according to Koch, Brierley, Maslin, and Lewis in their 2019 paper, *Earth system impacts of the European arrival and Great Dying in the Americas*

[526] Short 2024, page 96
[527] Flood 2010 revised edition, page 216

after 1492:

> ...multiple pathogens caused multiple waves of virgin soil epidemics over more than a century. Those who survived influenza, may later have succumbed to smallpox, while those who survived both, may then have caught a later wave of measles.[528]

A single disease would not have wiped out a tribe or a clan. But a disease that takes 30%, followed by another that takes 30% of whomever was left after the first epidemic, followed by another disease claiming 30%... These waves of illness added up to devastation, millions dead, and tremendous cultural and social loss that may not have been limited to the modern human indigenous, but also may have overwhelmed the relict hominins scattered among them, similarly defenseless to the European diseases.[529]

Smallpox was perhaps the worst of these diseases, transmitted through inhaled droplets from the coughing of the already infected or by physical contact. Part of the difficulty in containing an outbreak was due to the long incubation period of one to two weeks before the first symptoms of fever and body pains became obvious. In a couple of days a rash of red spots would spring up on the feet, hands, and face, spreading soon onto the body. The spots would grow larger, leaking fluids, itching, and appearing like blisters. If the afflicted lived long enough, the pox scabbed up, dried, and fell off leaving deep scars (pox marks). The disease would finally leave survivors about a month after infection.[530]

These epidemics left a largely emptied continental interior. Finding previous territory undefended, first displaced Native Americans, then white Europeans moved west into the under-inhabited and under-claimed territory. It wasn't until the 1930s that the Native American, Alaska Native, and First Nations people of North America stopped declining in population numbers and

---

[528] Koch et al 2019, page 22
[529] Boyd 1999
[530] Lange 2003

began to slowly increase.[531]

What we may be experiencing now—with increased Sasquatch sightings in the last 60 years—is an uncontacted indigenous culture gradually rebuilding in population and reestablishing its society and technologies in its old territories after a devastating period of decline driven mainly by introduced diseases.

In 1891, Patrick's future Skanicum father may have been an orphaned young teen, alone, clumsily doing his best to recover some semblance of family.

The Tasmanian culture also underwent a radical transformation when the Tasmanians were cut-off from the larger population of mainland Australia by rising sea levels. Joseph Henrich in his 2004 paper, *Demography and Cultural Evolution: How Adaptive Cultural Processes Can Produce Maladaptive Losses—The Tasmanian Case* (examining the reduction in Tasmanian technology during isolation), points out that:

> ...the relatively sudden reduction in the effective population size... could have initiated a cultural evolutionary process that (1) kept stable or even improved relatively simple technological skills, and (2) produced an increasing deterioration of more complex skills leading to the complete disappearance of some technologies and practices.[532]

It is exactly this type of de-evolution of technological skills that may have occurred to Sasquatch in North America. As a population's knowledge of its local environment expands, its need for technology to survive and thrive in that environment decreases. Sasquatch may have been more technologically equipped in the past—such as with fire, clothing, and spears—but as its intellectual understanding of place and resources increased, it's need for this excess baggage of gear and implements decreased. As a matter of efficiency, much knowledge may have been discarded over time.

Why carry a spear when a nearby rock will do?

---

[531] Koch et al 2019, page 22

[532] Henrich 2004, page 197

Sasquatch intelligence may actually have sharpened as their species came to rely less on things and more on sheer intellect to cope with environment. This was a process in contrast to the *self-domestication* of increasingly smaller brains that modern humans have been on for the last 20,000 years [see Chapter 16]. Instead, Sasquatch has been *re-wilding* through honing complex local environmental wisdom.

There are only a few reports indicating Sasquatch making or using tools. Early North American newspaper accounts occasionally referred to wild men carrying clubs or even wearing skins. One of Alley's witnesses reported a Southeast Alaskan Sasquatch digging for clams with a bat-sized stick.[533] The "Yowie" of Australia is similar or identical to the North American Sasquatch. One 1912 newspaper account from near Sydney, Australia was quoted by Jerome Clark in his 1993 book, *Encyclopedia of Strange and Unexplained Phenomena*, as describing an eyewitness seeing a drinking Yowie stand up and collect a stick from the ground next to it before it walked off.[534] Flood reports from the same continent that the Aboriginals used pointed stakes for digging. The 1912 Sydney account may have seen a valued digging-stick carried away to safety.[535]

As the old saying goes, "If you aren't looking for it, you won't find it." This may very well apply to tools associated with relict hominins. This applies even with *H. sapiens* populations of hunter-gatherers: "When early settlers found an Aboriginal tool that looked like a hoe, it was dismissed because they had convinced themselves that there was no agriculture in Australia." As Pascoe neatly puts it, "If you're not looking at these tools with an open mind, they are considered aberrations."[536] "...all of us must be alert to that greatest of limitations to wisdom: the assumption."[537]

Evidence of Sasquatch tool usage need not be limited to sticks and rocks. They may be capable of creating other technological

---

[533] Alley 2003 (2018 edition), pages 84-85
[534] Clark 1993 page 369
[535] Flood revised edition 2010, page 147
[536] Pascoe 2018, page 36
[537] Ibid., page 124

items noted among modern human hunter-gatherer groups such as containers or figurines out of clay, wood, animal skins, or intestines.[538]

Functional food-storage techniques in the wilds may be difficult for us to recognize. Pascoe points out that "many caches of grain and other foods were stored in grass packages, which were then smeared with clay to form an impenetrable vessel."[539] These packages may form a range of size and shapes, including spheres as small as a softball and incorporate preserving layers such as animal fats.

Alley had a singular report of two hunters mortally wounding a black bear on the Alaskan island of Revilla. The bear took refuge in a dense stand of trees in the middle of muskeg. The hunters thought it better to return the next day to find what they expected would be a dead black bear. But the black bear body was missing. After a long search in the open muskeg, they finally found the bear in a slight depression concealed with moss "patch-worked together, as if by hand." They left that area immediately and never returned.[540]

Although Sasquatch is often reported as throwing rocks, even in hunting, I can only recall one account of anyone finding rocks that appeared to have been worked by breaking or with controlled chipping. However, I once found a fist-sized rock on top of a boulder in the U.S. state of Colorado next to smashed bones (within sight of the lean-to stick structure pictured on page 237). If my suspicion is correct, I was seeing opportunistic tool use to break bones to get at the fat-rich marrow inside.

Tools aren't always necessary. Alley quotes a Prince of Wales Island witness as finding the remains of four deer with "every single long bone... cracked in half and the marrow sucked out!"[541]

Flood writes, "Stone tools played a very small part in traditional Aboriginal equipment: most artefacts were of wood, bone, shell, or plant material."[542] In Volume II of this series I will

---

538 Ibid., pages 145-147

539 Ibid., page 154

540 Alley 2003 (2018 edition), pages 232-233

541 Ibid., page 228

542 Pascoe 2018, page 146

introduce some baffling archaeological finds of worked rocks that may make more sense when looked at through Sasquatch-colored glasses. Perhaps finding Sasquatch tools made of stone is a long shot, but it is worth looking for other materials that, within their context, point to tool use.

Opportunistic tool use among Sasquatch is reported, though rarely. I recall one podcaster remarking that it was hard to hang onto five-gallon buckets in “booger country.” I recall another young teen describing how he heard and saw a juvenile Sasquatch using his missing axe and Alley describes a report in which an axe was taken from a campground.[543] Lapseritis lists apparent or suspected tool use including silverware and glass jars taken, a Sasquatch in Oregon seen carrying a large pot, and the occasional report of clubs in hand.[544]

An archaeologist for the U.S. National Park Service introduced me to the concept of constructed progressively narrowing chutes, potentially a mile or more long. In Alaska, these chutes, attributed to Alaska Natives, tended to be made on wide river bars. Low rock walls or even widely-spaced conceals or “hides” (constructions just big enough to hide hunters) were enough to steer the animals into a more and more confined route of attempted escape. She explained that the purpose of a chute was to guide game into a restricted kill zone. There have been a couple contemporary accounts of people finding themselves in deep woods in such chutes made of downed logs. Prey normally takes the path of least resistance and runs along such structures when in flight, rather than jumping over even low barriers like logs or rock walls.

It is possible that chutes thought to have been built by indigenous *H. sapiens* have actually been created by relict hominins, pointing to yet another area of potential research. Pascoe describes the remnants of such structures found in Australia:

> ...kilometres of brush fences in large scale trapping or battue operations. [Battues are both the process and the

[543] Alley 2003 (2018 edition), page 41
[544] Lapseritis 2011, page 21

structures that drive game toward hunters.] Remnants of the walls, outlining wings of these battues, can still be seen in some parts of the country.[545]

The following clam bed example from Pascoe is a technology that may be employed by the Yowie or the Forest People of North America.

> ...only in the last decade [the 2000s] was it discovered that Canadian First Nations peoples had been extending existing clam beds by building rock walls further away from the beach... The local First Nations people already knew of these clam 'gardens'... One further impediment to the revelation of this aspect of intensification was that the clam gardens were constructed and farmed by women and children—and such knowledge was never revealed to male archaeologists. Even after the engineering of the gardens had been examined by independent scientists, there was enormous reluctance to accept the results. The view seemed to be that such important structures could not have been overlooked by earlier archaeologists.[546]

Clam beds are just a start of what our more hairy friends could be capable of. In fish-rich areas we should also be on the lookout for other types of fish traps, such as what Pascoe describes in the quotes below:

> ...the Barragup fishing weir on the Serpentine River was described by Jesse Hammond circa 1860: 'A wicker fence was built across the stream, completely closing it from bank to bank, except in the centre, where a small opening was left.'[547]
>
> Uncle Max Harrison describes a vast fish trap in a bay near Bermagui. Now silted over, the trap comprised massive

[545] Pascoe 2018, pages 51-22
[546] Ibid., pages 81-82
[547] Ibid., page 69

> boulders...[548]

> Wiradjuri people in New South Wales also built large dams, and then carried fish and yabbies [a freshwater crayfish] in coolamons [wood or bark containers] over large distances to stock the new waterholes.[549]

In the western reaches of the huge Denali National Park, there are isolated kettle ponds. Kettle ponds are formed in the depressions left by huge chunks of melting glacial ice, making these ponds in Alaska less than 10,000 years old. These particular kettle ponds are not fed by or drain into other water systems. They are water islands in the thick muskeg. When these ponds occasionally fill from surface water, or overflow away, it is through thick mats of filtering vegetation—mosses and low shrubs. There are no open water courses. Still, some of these ponds are filled with Arctic Char, a salmonoid related to trout and salmon. The official explanation for their presence in these ponds is that eggs were stocked by some anonymous early European with a lot of time and resources on their hands. Never mind that there were no hatcheries within 500-miles when these fish stocks were found (and the nearest road still 100 miles away). *Sterile* Arctic Char are now reared in Fairbanks, but are not stocked at all into Denali National Park. We've already seen that Sasquatch can be quite territorial regarding fish stocks. Perhaps they have in the past transported fertile eggs to ponds, lakes, and streams in Alaska, Canada, and the Lower Forty-Eight. History might give Sasquatch a proprietary sense of ownership over such fish stocks.

"Aboriginal and Torres Strait Islander fishing systems were seen all over Australia in more or less complexity... not just functional devices, but also things of beauty."[550]

This next quote strikes me as particularly significant in understanding what at first glance seems like unnecessary aggression towards fishermen as reported in North America. "Particular ponds in the system were managed and used by

---

548 Ibid., page 70
549 Ibid., page 47
550 Ibid., page 87

particular families—but those families had responsibilities for the secure provision of fish to the families and systems upstream and downstream from their location."[551] This sense of responsibility for certain fish stocks at discrete locations might explain both the aggression towards *H. sapien* interlopers and a seeming sense of entitlement to help themselves to the catch in fish wheels, nets, stringers, drying racks, and even—occasionally—the holds of fishing boats.[552] Sasquatch may just be insisting on their senior fishing rights and be equally puzzled at our lack of understanding that *we* are the ones breaking the historic contract between groups.

Pascoe goes to great lengths to present material on the use of fire among Australian Aboriginals to shape their pre-European landscape and food supply. Essentially, large parts of Australia looked almost like British parks with isolated stands of trees breaking-up the swaths of tall grasslands that were themselves filled with game animals or even stretches of tuberous crops and grains.

> Aboriginal farmers had used fire to clear areas of land, which they were careful to separate with belts of timber... Aboriginals left the forest on poorer soils and cleared the best soils so they could create pastures and croplands. ...the explorer's journals suggest that colonial settlers ignored the Aboriginal method, and that contemporary Australians still suffer from the result.[553]

It would be interesting to investigate the possibility that our relict hominins once used fire in such a way in regions outside of Australia. A place to look for traces of this terra forming might be within old Native American, First Nations, and Alaska Native stories. The idea of the use of fire in pre-Columbian land management in North America has already been entertained. Citing earlier work by Doolittle 1992, Abrams and Nowacki 2008, and Anderson 2006—Koch, Brierley, Maslin, and Lewis in their

---

[551] Ibid., page 75
[552] Alley 2003, page 51
[553] Pascoe 2018, page 24

2019 paper, *Earth system impacts of the European arrival and Great Dying in the Americas after 1492*, contend fire had been used both for slash and burn agriculture and for, "...large-scale fire-based forest clearing to support indigenous hunting strategies."[554] These were two fire goals also used pre-contact by the Australian Aborigines.

What we assume was virgin forest when Columbus approached North America, probably was not virgin forest at all. In the same paper referenced above, the authors suggest that, pre-European contact, most of New England was under extensive land use and the remaining land east of the Mississippi was under the lower standard, *intensive* land use.[555] So why this impression that vast unbroken forest greeted the Pilgrims from their vantage of Plymouth Rock in 1620? Because by 1600, as many as 90% of North American Native Peoples had *already* died from the diseases that spread quickly after first contact a hundred years before, racing in front of European exploration and settlement,[556] literally making Anglo occupation of North and South America possible. Lacking the labor to keep forests at bay, secondary succession of woodlands on formerly open country would, within 100 years, appear much like undisturbed old-growth forests.[557] By the time settlers were moving into the interior of New England and points further south, it looked like they were advancing into the forest primeval. Traces of previous land management by Native Americans, and possibly Sasquatch, would not have been expected, looked for, or reported.

This springing-up of new forest on previously open lands apparently had a global impact. Koch et all, credit the re-growth of forest on an estimated 138 million acres of land in the Americas as sequestering sufficient carbon to nudge the world into the Little Ice Age, with three separate cooling intervals between 1650 and 1850.[558]

Another simple trace that Sasquatch or Aboriginal presence

---

[554] Koch et al 2019, pages 19-20
[555] Ibid., Figure 2 page 16
[556] Ibid., page 13
[557] Ibid., page 23
[558] Ibid., pages 13 and 30

would be expected to leave on the landscape is that of trails. On one of my Alaskan wilderness hikes, I found a wide and freshly disturbed trail in the woods 100-feet back from an open creek bed. This was in a V-shaped valley known as "Bear Draw". I had left the pleasant rocky creek bed behind me and was searching dense white spruce woods for an old cabin I had heard about. This trail paralleled the creek but petered out in *both* directions. It was in an excellent position from which to view anything traveling on the creek. I was already aware of the concept of a "Big Boy trail" and became alert when I noticed that this churned-up path avoided higher branches on flanking trees that bears would just go under. This trail was obviously made by something at least as tall as I. A torn-up rotten stump and a freshly broken willow trunk completed my impression that this trail had potential origins of interest. In the same area I have also encountered a loud and deep call from the woods ("Hoy!") and fiber-filled, but human-looking scat. I believe that anomalous trails are excellent trace evidence of Sasquatch presence, but be careful if you find such a trail: You may be straying into an area where you are not welcome.

Sasquatch reports are generally very consistent in terms of shelters, ranges, and technologies across thousands of miles. I already mentioned Flood's point that Aboriginal tribes formed a "complex social and economic network" that transported goods, but also moved ideas about the continent.[559] Wiitala had a novel observation that is worth noting. What he found was a suspected rodent-trap made of rocks. He was vague about its appearance and construction, but he described a pile of rocks on an old logging road that he thought may have been assembled to create shelter for rodents that could be quickly opened to nab the furry calories hiding inside.[560] Alley reported an interesting theory from Fred Bradshaw: The rock stacks may be designed to trap the sun's warmth, providing a double incentive for rodents to shelter there.[561] The core idea behind piles of rocks could propagate quickly across North America (if it is in fact new): Make suitable

559 Flood revised edition 2010, page 268
560 Wiitala 2021 page 44
561 Alley 2003 (2018 edition), page 257

habitat to attract prey species that can be turned into a trap for those sheltering animals. These traps could be assembled out of rocks, boulders, branches, brush, dirt, or logs on land, in trees, or even in water (for fish and shellfish). These are structures that previously may have appeared incompressible, irrelevant, natural, or random. But now Wiitala has suggested a purpose and a creator.

Flood, in referring to a time when the world oceans rose with the melting of the polar ice caps, wrote, "The tremendous loss of territory, particularly if it happened fairly rapidly, may well have been what triggered migration and conflict…"

A similar shock wave might have run through the Sasquatch population starting in 1492. A comparatively gigantic, but peaceful and shy branch of humanity may have gone through enormous changes in culture and technology as they adapted to this new threat of the pale Anglos. "New weapons tend to spread faster than other artefacts." [562] The new "weapons" of Sasquatch were not spear-throwers or bows. The new weapons were stealth, hyper-vigilance, coded inter-group communications (whistles, animal calls, and knocks), and determined concealment to protect their children and core strongholds.

[562] Flood revised edition 2010, page 268

Before Patty

CHAPTER TWENTY

Where Did Sasquatch Come From?

"Two nations are in your womb,
And two peoples within you will be separated;
One people will be stronger than the other,
and the older will serve the younger."

---*The Old Testament*; Genesis 25: 23

I lived in western Massachusetts in 2024 and 2025. On grocery trips to a nearby town, I often passed a trailhead identified as "Esau's Heel." This curious trail name caught my eye and stirred an old memory from a brush with Lutheran church services in my youth. As the Old Testament Biblical account goes, Esau—strong, red, and hairy (red-haired?)—was the first of twins born in 1836 BC to Rebekah, who was married to Isaac, son of Abraham. Jacob was the other twin. There were several locations in the immediate area of Esau's Heel Trail named after Jacob.

I eventually stopped to walk the gentle trail. I was not expecting to see anything related to Sasquatch, but I was wondering how the trail got its name. Was it possible that large tracks had been found in the area and the trail name was literally the reference to heel impressions from a hairy creature or was it just a literary reference to a Bible story?

To my bemusement, I soon found anomalous tree breaks along this frequently used trail. In the first half mile, many young white pine trees had broken tops eight to ten feet (2.5 – 3.0 m) off the ground (as pictured below). Where snapped, the trunks were approximately two to three inches in diameter (5 - 8 cm). While ice storms are a possible explanation for the damage, the more brittle deciduous trees growing along the trail did not show similar breaks. And wind damage was not a good explanation because these were young pines, protected from heavy winds by the much taller trees sheltering them.

Sign at the trailhead to Esau's Heel trail in western Massachusetts, USA

Perhaps coincidentally, within the controversial 2012/2013 Melba Ketchum et al genetic sequencing study of purported Sasquatch DNA from hair, tissue, and blood samples—*Novel North American Hominins, Next Generation Sequencing of Three Whole Genomes and Associated Studies*—some of the modern human haplotypes identified on the maternal (mitochondrial) side were of Middle Eastern origin.[563] Haplotypes are specific and identifiable bundles of genetic material inherited from a single ancestor and may be tens of thousands of years old (or more than one million years old as is the case with human haplotype X).

December 2024 tree break in a young white pine along the Esau's Heel trail in Massachusetts

The Esau and Jacob story has some intriguing hints possibly relevant to the Sasquatch origin mystery. The epigraph that began this chapter were the words of the Lord speaking to Rebekah, a pregnant woman expecting twins.

"**Two nations are in your womb**…" "Nations" may be translated as peoples. Two distinct peoples growing in the womb of the *Homo sapiens* woman, Rebekah. Genesis 25: 24 expands:

---

[563] Ketchum et al 2013, page 15

"The first to come out was red, and his whole body was like a hairy garment; so they named him Esau." "Esau" is thought to mean "hairy." Easau, in body, must have seemed to represent a distinct nation. Patrick also was two nations within Hahissiat's womb. In Volume II of this series, I will consider the possibility that we (modern *Homo sapiens* and Sasquatch) are literally two nations that came from the same womb; brothers parted.

"**And two peoples within you will be separated**…" Although brothers of a single mother, the two sons took different paths and were separated from each other. We are separated still.

"**One people will be stronger than the other**…" This is the one overt reference in Genesis 25: 19-34 to Esau's physical strength. We are told in verse 27 that Esau was a skilled hunter and "a man of the open country." "Open" in this case likely means country that is not claimed as property—the wilds.

"…**the older will serve the younger**." As the story goes, Jacob was born clutching the heel of his older brother and the name Jacob translates as "he grasps the heel." Grasping the heel has become an expression for deception and a synonym for a heel is a cad. Sadly, this is the story of the younger brother, Jacob, cheating the elder Esau out of his inheritance birthright. Esau lived in the moment and traded away his birthright to the scheming Jacob in exchange for lentil stew and bread when Esau was very hungry.

If these four lines contain the germ of the greater story between the naïve elder brother—Sasquatch—and the scheming younger brother—Caucasians—is it any wonder that Skanicum distrusts us still with our camera traps, high-powered rifles, gifting stations, and call blasting?

Another interesting sidenote: Some North American tribes, such as the Lakota and Cree, call Sasquatch the Elder Brother.[564]

A Native American perspective of the origins of Sasquatch comes to us from the California Yurok Indian, Lucy Thompson, from 1916:

> Our Indian devils are Indians who for some reason or cause leave the tribe and go far away into the lonely mountains and

---

[564] Short 2024, pages 57 & 95

> into the depths of the forests, where they live near the streams and places almost inaccessible. In their loneliness, they roam through the forests and over the mountains like some wild animals of prey. They forget the language of their mothers and become something like wild beasts, fleeing from the sight of human beings.[565]

In this tradition, members of the *H. sapiens* family isolate themselves and in some individual transformation, change physically to fit their new, wild circumstances, becoming "Stick Indians." Although this de-evolution sounds fanciful, it is another theme I will explore seriously in Volume II.

Pyle echoes this concept of transmutation to meet the needs of the deep forest when he writes, "...Sasquatch is a recognition of the connection between the human and the natural."[566] Pyle also explores western mythology to find many potential representations of male and female Sasquatch in Pan, Puck, Silenus, Dionysus, Enkidu of Gilgamesh, Diana, Demeter, Astarte, Gaia, Maya, the Great Goddess, and even Mother Nature.[567]

This confusing array of echoes of Sasquatch should remind us that we are not necessarily grappling with just one "breed" or even one species. There are reports of hairy bipeds from North and South America, Europe, Asia, Africa, and Australia in a variety of sizes (to me, four distinct size ranges), habitats, morphology, and cultural sophistication. We may be dealing with several genetically distinct species. Lapseritis claims that the Sasquatch in Indiana self-report (presumably through telepathy) that there are two kinds of Sasquatch in that immediate area, able to interbreed with each other. One type looks very human-like, only hairy and much bigger. The second type looks very gorilla-like and the adult males may only reach a little over 6' tall (2 m).[568] Paulides (through witness testimony rather than telepathy) similarly identifies two strains of relict hominin the classic

[565] Thompson 1916, page 129
[566] Pyle 1995, 2017 edition, page 188
[567] Ibid., page 186
[568] Lapseritis 1998, page 49

Patty-type of "Bigfoot" and another that resembles a "wild man"—Thompson's Stick Indians?

Researchers have picked through evolutionary research and findings for traces or leads for the origins of Sasquatch. The large Asian ape (?), *Gigantopithecus blacki*, is often pointed to as a possible fossil predecessor to modern Sasquatch, with recovered remains as old as 2 million years. Certainly the enormous estimated size of *Gigantopithecus* checks one of the boxes. Unfortunately, only teeth and four mandibles (jaws) have been found so far—insufficient to claim bipedalism for the species. Many sources insist that *G. blacki* died out 300,000 years ago, but Lopatin, Mashenko, and Le Xuan Dac published the 2021 paper, *Gigantopithecus blacki (Primates, Ponginae) from the Lang Trang Cave (Northern Vietnam): The Latest Gigantopithecus in the Late Pleistocene?*, suggesting that *G. blacki* persisted until at least 125,000 years ago.[569] This late Pleistocene dating means they would have been coexistent in time, *but not necessarily in locale*, with various *Homo* line species. The current list of the generally accepted *Homo* species (from oldest to the most recent): *Habilis*, *georgicus*, *ergaster*, *antecessor*, *erectus*, *heidelbergensis*, *naledi*, *floresiensis*, *neanderthalensis*, and *sapiens*. *Homo juluensis*, the "Big Heads" may be joining this list soon. *H. juluensis* may include the Denisovans.

According to the accepted archaeological record, *H. sapiens* and *Gigantopithecus* probably did not meet each other (although failed early Out-of-Africa migrations of *H. sapiens* may have), but it is accepted that *H. erectus*, surviving in Asia until 30,000 years ago, almost certainly did encounter *Gigantopithecus*.[570] *H. erectus* skulls share some features with Patrick: a sloping and low forehead, weak chin, and a long and low cranial vault.[571] What would the result have looked like if the human-sized *H. erectus* interbred with *Gigantopithecus*?

This speculation leads to the issue of hybrids: Crosses between

---

[569] Lopatin et al 2021, page 4
[570] Debenat 2009, Page 179
[571] Roberts et al 2018 edition, page 125

similar species that may or may not prove to be viable breeders themselves. Patrick was a hybrid and he was obviously viable. Paulides points to past and possibly ongoing hybridization with native peoples when he writes, "We believe this biped is far from an ape, far more intelligent, and possibly aligned genetically with the Native North American."[572] Leo Frank, in his 2022 book, *Sasquatch: Family Ties*, writes that his personal experience suggests that different varieties of relict hominin are interbreeding and produce hybrids of their own with a great deal of varied morphologies as a result.[573] This mixing may contribute to "hybrid vigor" (heterosis) in the populations and explain the confusing variation of witness reports that range from descriptions of huge bipedal orangutans to thinly-haired cave men. Hybrid vigor may lead to enhanced capacities in such traits as physique, health, or intelligence gained from having parents with differing, but compatible genetic identities. One theory to explain this hybrid vigor is that genes such as those for sickle cell anemia, hemophilia, cystic fibrosis, and color blindness are not able to *express*—become physically present—when they do not find a matching recessive gene in the other parent's genetic material.

Before leaving this discussion of genes and Sasquatch that is perhaps more relevant to Volume II or the next chapter on hybrids, it may be appropriate to consider another potentially genetically transmitted trait affecting both Sasquatch and *H. sapiens*—fear of each other. Redfern in 2016 touched upon this fear[574] and it is a theme throughout Vendramini's 2009 work. Vendramini explains the effects of the genetic transmission of fear through chronic-stress, triggering DNA methylation (a chemical process that may turn genes on or off). "Even if a rat only smells a cat, its brain releases a cocktail of potent stress response hormones that renders it anxious and hypervigilant."[575] Isdahl also notes the puzzling witness tendency to not speak of encounter events even among people that shared the sighting. In some cases these witnesses

[572] Paulides 2017, page 465
[573] Frank 2021, p 189
[574] Redfern 2016, pages 14-16
[575] Vendramini 2009, page 97

even stop speaking to each other entirely. A genetically-imprinted fear—a reaction noted frequently in both modern humans and Sasquatch—implies thousands of years of conflict between the two "brothers." We are afraid of each other because many times over thousands and thousands of years we have reinforced the good sense and practicality of that fear. It has thus far been wise to steer clear of each other.

It is frustrating that Sasquatch *apparently* does not show itself in the fossil record. Such a big animal should leave big fossils, yes? Maybe not, actually. Pyle points out that gorillas and chimps are entirely absent from this record.[576] And those apes are pretty big and have been around for a very long time.

However, genetic approaches are beginning to ferret out mysteries of evolution even without hard fossil evidence. Genetic codes leave a record of every ancestor that contributed to the individual's makeup. We can now determine how much, if any, Neanderthal or Denisovan blood flows through us. And remember, we haven't even assigned a genus and species name to the Denisovans. It is established now that there is a third unclassified hominin that contributed to modern human genes. We have not found a scrap of fossil evidence for this third brother, but it lurks, concealed, in our genetic code. This "ghost progenitor" may even be our elder brother Sasquatch. It is likely that there will be additional hominins found sheltering in our genes.

Pyle continues by pointing out that the *Homo* line diverged from *Pan* (the chimpanzees) about 8 million years ago and from *Gorilla* about 12 million years ago. Even with these millions of years separating us, we share a 97.7% genetic match with gorillas and an uncomfortable 98.4% with chimps.[577] We see such close family ties within the occasional re-emergence of old physical traits such as excessive hairiness per Marlowe:

> ...researchers have postulated that CGH [congenital generalized hypertrichosis] is a manifestation of a genetic

---

[576] Pyle 1995, 2017 edition, page 385
[577] Ibid., page 386

> atavism—reappearance of an ancestral phenotype. The reappearance of ancestral characteristics in individual members of our species…[578]

A phenotype is the set of observable physical traits that an organism develops through the interaction of its genes with its environment. If we interacted with the same environment as Skanicum, would we appear so different from each other?

"Reappearance of an ancestral phenotype" is a useful way to describe the world-wide phenomenon of Sasquatch returning to us from our past. We are not so very far removed from Patrick or even from his Skanicum father.

[578] Marlowe 2013 page 83

CHAPTER TWENTY-ONE

Hybrids

"...a discrete population of organisms
that doesn't naturally interbreed with other groups."

---The definition of "species" from
Evolution: The Human Story;
Dr. Alice Roberts, et al, 2018 edition, page 37.

Hybridization is simply when one species successfully breeds with another genetically very similar species.

There are a few factors which may lead to hybridization attempts, including: 1) Opportunity; 2) Reproductive failure within a species (such as with infertility in one gender, the unavailability of suitable mates, or due to genetic isolation—the lack of locally-available sufficiently genetically-distant mates); 3) New contact between formerly separated species (such as with expanding home ranges); and 4) When hybridization confers a reproductive advantage—such as the advantage of enhanced survivability of young.

Normally in mammals, fertile hybrid offspring may only occur between species with the same number of nuclear chromosomes (those contained within the cell nucleus). Humans have 23 pairs or 46 total chromosomes. Chromosomes are complex molecules of protein, arranged in pairs, that carry the genetic information of the individual and species, inherited equally from the mother and the father. In an example of an infertile hybridization, a horse with 64 chromosomes may mate with a donkey with 62 chromosomes to produce a normally infertile mule with 63 chromosomes.

In 2021, an odd-looking dog was found in Brazil. Genetic testing revealed that this "dogxim" was the product of a union between a pampas fox (with 74 chromosomes) and a domestic dog (with 78 chromosomes). The dogxim had 76 chromosomes, but was probably infertile (it died in 2023). Significantly, foxes and the *canis* (dog) group diverged evolutionarily 6.7 million years ago, demonstrating just how far species may drift and still be able to produce offspring, though in this case the offspring was non-viable.[579]

The effect of a successful and viable hybridization may be that of an *introgression* of genetic material into one of the species involved. Patrick's Skanicum father created a genetic introgression into modern humans. Later chapters will follow-up on Patrick's success in conveying his and his father's genes into North American *H. sapiens*.

Hybridizations may involve gender-dependent *growth dysplasia* in the offspring. Growth dysplasia in this case refers to abnormal size in hybrid offspring. This can result in offspring either substantially bigger or *smaller* than either parent. Many species, but not all, have male genetics that tend towards maximizing the size of offspring in order to maximize their reproductive success in competition with other males. But females of the same species that have young by multiple males, such as lionesses, carry growth-limiting genes on the female side. These growth-limiting genes, in the case of lions, are thought to exist to cancel or dampen the effects of the growth-maximizing genes of the male lions that she has mated with. Her offspring then remain within a given size range, without unusually large males killing off their half-brothers and reducing the lioness' own reproductive

---

[579] 9/27/2023 https://www.nationalgeographic.com/animals/article/dogxim-first-dog-fox-hybrid [accessed 2026.01.25]

success.

It appears that tigresses do not have these growth-limiting genes so that when a female tiger (up to about 350 lbs or 160 kg) mates with a male lion (up to about 600 lbs or 270 kg), the male lion's growth-maximizing genes have the opportunity to fully express. The result of such a hybridization may be a male **Liger** weighing up to 900 pounds (410 kg)! Tigers and lions (and most cat species) each have 38 chromosomes so the offspring of matings between tigers and lions are generally fertile.

But when a male tiger (up to about 450 lbs or 200 kg) mates with a female lion (up to 400 lbs or 180 kg) the result is a **Tigon** that weighs no more than the lioness. As it turns out, male tigers carry the growth-limiting genes themselves.

Fortunately, many animals have growth-limiting genes that act specifically on natal size. A female may give birth to a normal-sized infant that then goes on to grow into a relative giant, such as a Liger. Dog breed mixes frequently have offspring that are bigger than either parent. A witness on *Sasquatch Chronicles* episode 936 mentioned that a European Wild Boar mating with a feral North American hog will produce much bigger offspring than either parent.

Within primates, baboon hybrids often involve increases in size from both parent species as part of their hybrid vigor. Other potential outcomes of hybridizations include larger body size, larger faces, and unexpected dentistry including larger and extra teeth.[580]

Perhaps Sasquatch is itself a product of hybridization involving a lack of growth-limiting genes in the presence of growth-maximizing genes. If Sasquatch is a hybridization of two or even more species, it may be bigger than any of the species that created it.

The occurrence of hybridizations are established fact within the *Homo* line. Higham discusses how a bone fragment found in Denisova cave in Siberia showed, through ZooMS (zooarchaeology by mass spectrometry), that the owner of that bone fragment had a Neanderthal mother and a Denisovan

---

[580] Higham 2021, page 234

father.[581] Not only was this a first generation hybrid, but the remains of "Denny" found in Denisova cave ultimately revealed that her Denisovan father himself had genetic traces of other Neanderthal admixtures from *hundreds* of generations before.[582]

It should be noted that the Denisovans have not been widely identified as a specific *Homo* species, mainly due to a lack of sufficient skeletal remains (such as a skull) to provide a characteristic morphology (the type specimen). Denisovans are thought of as an unidentified human population, genetically distinct from *Homo sapiens.* However, there have been intriguing fossil remains discovered further east in China that make a case for taxonomic identification as a distinct species which *may* include the Denisovan population. These "Big Heads" (due to their large cranial capacity) have been dubbed *Homo juluensis*[583] and will be included in discussions in Volume II.

Patrick was another successful hybridization event. That success alone implies that his Skanicum father was sufficiently closely related to *H. sapiens* as to ensure his classification within the *Homo* genus.

---

[581] Ibid., page 1
[582] Ibid., pages 110-111
[583] Bae 2024

CHAPTER TWENTY-TWO

Hybrids in the Human Family

"Evidence shows that these different groups of humans regularly interbred when they met… There are major implications for our traditional concept of 'species' and what a species is."

—Tom Higham, *The World Before Us: The New Science Behind Our Human Origins*, 2021, page 223

This chapter will look more closely at the likelihood of genetic introgressions into *Homo sapiens* through hybridization events such as what occurred with the birth of Patrick. In the epigraph above, Higham suggests that the concept of species may need refining. Interbreeding may not be commonplace, but also not necessarily rare as Higham pointed out. He continues his explanation:

> ...there is an increasingly long list of species that regularly hybridize with others. Baboons are an interesting primate

example that offers a sobering parallel to the genus *Homo*.[584]

Higham writes that hybridizations among primates have occurred in species that have diverged as much as four million years previously.[585] Humans are included in the mammal order of Primates and are not immune to the occasional pressures that lead to hybridizations.

The tailless great apes split off from monkeys about 30 million years ago. Gibbons then split off from the great apes by 20 million years ago, leaving the great ape family Hominidae which includes, *Homo* (humans), the panins (of the genus *Pan*: chimpanzees, Bili apes, and bonobos), gorillas, and orangutans.

Humans have 46 chromosomes, while the other great apes have 48. The chromosomes of the one will not pair properly with the other, short of infertile "mules" or some inadvisable chimeric gene-splicing laboratory work. One might think that the essential chromosomal mismatch would have settled the possibility of *Homo* hybridizing with any of the other great apes, but there are persistent rumors that hybridizations between humans and chimpanzees have been attempted. These rumors seem horrifyingly factual when it comes to the unethical work of the Russian, Ilya Ivanov, in the early twentieth century (before the current level of DNA understanding). Ivanov's diabolical work (thoroughly documented by Redfern) was carried out in secret labs in places as far flung as French Guiana and Eurasian Georgia, through various avenues of experiment, all ending in failure. Ivanov was eventually arrested by Soviet authorities on New Year's Eve 1932 and died in exile two years after his sentencing.[586]

Apparently we need to look for potential hybrid matches from relatives more recently related to us than that of our approximate 7-million-year-old chimpanzee connection. Even so, Higham writes that there is, "...much emerging evidence for interbreeding [among *Homo* species]."[587] Later, he notes that an Oxford team

---

[584] Higham 2021, page 233
[585] Ibid., page 46
[586] Redfern 2016, pages 131-133
[587] Higham 2021, page 157

"...was able to show that there would have been *no reduction in fertility* between humans, Neanderthals, Denisovans, or even with older species of our genus such as *Homo heidelbergensis*."[588] There is some debate if *H. sapiens* evolved out of *heidelbergensis*, but if indeed we did, the split was no later than 300,000 years ago (when we appear to have lost touch with the last disputed *heidelbergensis* holdouts in Africa). Apparently a 300,000-year gap is no barrier to a fruitful reunion among hominins. You may recall that Lopatin, Mashenko, and Le Xuan Dac believe that *Gigantopithecus blacki* was alive and well in southeast Asia as late as 125,000 years ago.[589] There are, of course, a few in the Sasquatch world that suggest that *G. blacki* might have been an erect biped, possibly even of the *Homo* line. If so, past hybridization events with that species would be conceivable. Remember, it is established that the earth had no lack of *Homo* species from 300,000 years ago up to about 30,000 years ago, including, during some or all of that time: *habilis*, *georgicus*, *ergaster*, *antecessor*, *erectus*, *heidelbergensis*, *naledi*, *floresiensis, juluensis* [if the "Big Heads" are eventually accepted as a distinct species], *neanderthalensis*, and *sapiens*. According to Higham, "...there may have been more than three overlapping human groups in Africa 200-300,000 years ago."[590]

Not all of the home ranges of these *Homo* variations overlapped, but still, the potential for hybrids over that many thousands of years among that many *Homo* species is enormous and may explain some of the peculiar anthropological finds of anomalous skeletal features.

Higham sums it up: "...the most parsimonious reading of the accumulated evidence thus far is for a mosaic of human groups, geographically separated but meeting and interbreeding occasionally."[591] This occasional interbreeding would leave behind a trail of genetic introgressions, with shared genetic traits and physical features. Higham further narrows down a period which saw significant numbers of offshoots in the human family:

---

588 Ibid., pages 235-236

589 Lopatin et al 2021, page 4

590 Higham 2021, page 17

591 Ibid., page 237

> ...the Earth was a primevally complicated place 50-100,000 years ago, a Middle Earth in which the human family comprised a multitude of different types and sizes, living in a variety of ecosystems. This variety occurred owing to the isolation of populations from one another, which set them on a divergent evolutionary path.[592]

This potential for genetic introgressions, *back-and-forth*, is undeniably evident in our own genes as Higham points out: "...around 7.5 percent of the genome of the Melanesian people in all probability derive from population admixture with these now disappeared human groups." [2.6 % from Neanderthals and 4.8 % from the Denisovans] These introgressions make for confusing and complex interrelationships within our growing understanding of the human family: "The branchy tree of human evolution should instead be replaced with a wide, braided river, whose tributaries often flow into and away from one another."

Genetic studies show that *H. sapiens* and Neanderthals overlapped in Europe for at least 2,500 years after first bumping into each other in the African break-out lands of the Levant. The Levant lies in the western Middle East and is now occupied by all or parts of Israel, Jordan, Lebanon, Syria, and southern Turkey. Genetic testing of a large femur found in Siberia point to a Neanderthal introgression into modern humans about 55,000 years ago.[593] Genetic studies indicate that this admixture into *H. sapiens* came from one genetically homogenous group of Neanderthals.[594] That genetic similarity may imply a Neanderthal population bottleneck (sustained low population size), one of the effective motivators to hybridize.

Vendramini suggests a predatory aspect to this genetic introgression—cannibalism and rape—and a blurring of exterior form:

---

592 Ibid., page 172
593 Ibid., pages 142-145
594 Ibid., page 180

> ...the period of predation resulted in numbers of hybridised half-Neanderthal/half humans. In theory, these hybrids would have features from both species, and eventually this would affect the appearance of each population.[595]

Denisovan admixture happened even more recently than the Neanderthal introgression (Higham 2021).[596] Denisovan traces peak at about 4.8% of the genes in Melanesian Papuans and Aboriginal Australians. Interestingly (at least to me), I have measurable amounts of Melanesian genes according to my own genetic testing. This testing was done before they were identifying Denisovan genes as a separate category. Although I am very much a sole product of northern Europe, I suspect this trace of Melanesian is due to another Denisovan introgression into what are now Scandinavian populations. I presume this admixture comes through my Finnish genes. The Finns' genetic make-up has been influenced by migrations from East Eurasian Siberia, as indicated by the presence of the N1C haplogroup. I suspect Denisovan admixture (the measurable trace identified as Melanesian) may have been carried by those East Eurasians and associated with, though not necessarily derived from, N1C.[597]

In total, there were at least two, perhaps three or more Denisovan genetic introgressions into *H. sapiens* from groups of Denisovans "genetically divergent" *from each other* (Higham 2021).[598] New Guinea, Australia, and Tasmania were populated by *H. sapiens* from the east, across the Wallace Line (a minimum 80-mile crossing of deep ocean waters—a complete barrier to most animal species)—while the Sahul was still one large land mass exposed during the last Ice Age, a period of much lower sea levels. After the post Ice Age sea level rise, the Sahul was broken into the smaller land masses of New Guinea, Australia, and Tasmania. Higham believes that the Denisovan introgression

---

[595] Vendramini 2009, page 104

[596] Higham 2021, page 188

[597] Note: The N1C haplogroup was **not** one of those found by Ketchum et al 2013. Interestingly, the mitochondrial and nuclear DNA sequenced showed no Neanderthal or Denisovan sequences either. Pages 13-14, 33.

[598] Ibid., pages 179-180 and 182

happened while there was a single ancestral *H. sapiens* population—before they crossed the deep waters of the Wallace Line and dispersed into the Sahul. He feels this introgression probably happened within island Southeast Asia. To Higham, this presence in the islands of southeast Asia implies that the Denisovans, "...must have been able to adapt to very different environments and had a considerable technological and adaptive ability to survive."[599]

H. sapiens, along with the rest of the arbitrarily defined *Homo* line, appears more and more to be a multi-hybridized mongrel species. Hominins may not appear any more closely related to each other than a Siberian Husky and an Irish wolf hound, but those two breeds have no problem, given the opportunity, to produce a litter of puppies with a huge variation in physical form. As we look closer and closer at the many shapes of humanity, the idea of *species*—as a unique label—becomes a meaningless distinction. Whatever Skanicum is, I believe it enjoys a place in the pantheon of human breeds and should be recognized as such.

Anthropology supports this long view of repeated hybridizations with evidence of multiple hominins scattered across Africa and Eurasia:

> Scholars have suggested that there were probably phases of isolation and independent evolution happening in different regions in Africa, with periodic interbreeding and contact, and then at some point between 100,000 and 150,000 years ago the emergence of a population or populations of *Homo sapiens* with a relatively modern appearance… The picture is made even more complex by the high chance that this diversity in human groups was magnified by the presence of other *types* of humans in Africa too.[600]

It is commonly supposed that the earliest migrations out of Africa faded out of existence (or perhaps were absorbed into other *Homo* lines?) before a final successful outmigration no earlier

---

[599] Ibid., pages 175-176
[600] Ibid., pages 14-15

than 80,000 years ago. However, there appear to be traces of earlier migrations within the genes of scattered southeast Asian Negritos (the Mamanwa of the Philippines, the Jehai of Malaysia, and the Onge of the Andaman Islands). Higham writes of the Negritos that they, "...are characteristically small in height, with a gracile build and dark pigmentation... They may represent a relict population from an early out-of-Africa diaspora."[601]

Genetic studies of these old populations, such as published by Larena et al in 2021, find interesting evidence of *Homo* line hybridization events. Denisovan admixture or introgression is the highest in the world among the Ayta Magbukon Negritos, from the center of the Luzon Island in the Philippines, at up to 40% higher than among the Papuans (approximately 6-1/2%). The many authors of this study suggest that, "...Denisovans were probably widespread among ISEA [Island Southeast Asia]..." The authors believe that the Negritos and the Papuans experienced final Denisovan introgressions as recently as 25,000 years ago *at separate locations* in ISEA. The authors come to some startling conclusions based on their data, not the least of which is, "the possibility that the Denisovans comprised deeply structured populations with considerable genetic and phenotypic [physical form] diversity, enabling them to adapt into a wide variety of environments and thus inhabit a broad geographic range across the Asia-Pacific region."[602]

Loosely paraphrased: The Denisovans had a huge range, they were hybridized *Homo* mutts with a large range in their appearance, and they were very adaptable. If the Denisovan were indeed the "Big Heads" (*H. juluensis*) we may be able to add highly intelligent to this list.

Roberts pointed out in 2018 that, "The DNA of modern Africans contains traces of other interbreeding events, too, with as yet unknown archaic human populations."[603] Higham produces more recent evidence in this rapidly evolving study of the human family:

---

[601] Higham 2021, page 169
[602] Larena et al 2021
[603] Roberts 2018 edition, page 191

In 2020 scientists reported finding DNA in modern African people that could not be sequenced to any other known human genome. This is taken as possible evidence of a "ghost" population that must have lived in Africa tens or hundreds of thousands of years ago.[604]

This ghost population in Africa might help to explain the puzzling results of Zana's son's genetic tests. Zana was a supposed captive Russian Almas (the Eurasian version of Sasquatch) in the 1800s. The genetic tests of her son's exhumed remains indicate an African heritage. Zana may have been a hybrid between this ghost population and African *H. sapiens*, whose forbears passed through the Levant into Eurasia. If so, this ghost population may have been closely related to the Almas and even to Sasquatch.

Higham suggests that it is possible that an "archaic genome" found in testing for Neanderthal and Denisovan genes, may come "...from another archaic hominin, as yet unidentified genetically, that also introgressed into our ancestral human genome thousands of years ago."[605] Presumably, this introgression occurred in Eurasia, the known home of both Neanderthals and Denisovans.

Tim "Coonbo" Baker insists that he has personally met humans that have interbred with Sasquatch. It was unclear if he meant that he had met the progeny of such unions or if these were people who claimed to have engaged in such unions that were productive.[606]

Higham poetically concludes, "We are not simply human, we are the sum of all the branches of that braided river that touched and parted on our way to today."[607]

There are indications that Sasquatch is a hybrid population. Debenat suspects that the potential gene pool is complicated in that there are likely multiple relict hominid bipeds extant on the Earth right now. Some of these hominids are sufficiently closely related to *H. sapiens* for hybridization, but some are likely not

[604] Higham 2021, pages 16-17
[605] Ibid., page 180
[606] Baker 2005
[607] Higham 2021, page 239

close enough relatives.[608] Pyle describes a conversation with an old timer in Washington state, "...he told me with some urgency that some of the Indians feel, and he shares their belief, that there may be several types of Bigfoot… and that they may interbreed."[609] And Frank feels that some of the phenomenon in North America is a Sasquatch/Neanderthal hybrid.[610]

From Igor Burtsev, the esteemed bargaining Russian researcher:

> "...there is no indication that bigfoots are the result of parallel evolution and only distant relatives of humans. On the contrary, there are signs of a very close relationship. If so, crossbreeding can be banked on. In Europe, Asia and Australia there are legends, as well as old and not so old reports, of crossbreeding between 'wildmen' and normal humans.[611]

Ronald Morehead, in my 2012 Kindle edition of *Voices in the Wilderness: A True Story*, mentions the possibility of hybrids in the introduction.[612] Alley writes that "there may be two similar large species coexisting besides man in North America and elsewhere, even possibly hybridizing [with each other]."[613] And Bland confidently asserts that Sasquatch is a human hybrid.[614]

Clearly, the main premise of this book is that Patrick was a Sasquatch/Human hybrid. If Sasquatch itself is a hybrid of two or more hominins, predicting the physical outcome of such a mating is moving into uncertain territory. It is not accurate to think of the size of the product of a mating to be the average of the contributing father and mother. A very tall and heavily built Skanicum mating with a "tiny" human woman (as Hahissiat was described) is not going to produce a medium-sized Skanicum because female *Homo sapiens* carry growth limiting genes, much

---

[608] Debenat 2009
[609] Pyle 1995, 2017 edition, page 344-351
[610] Frank, 2021, page 87
[611] Burtsev 2015 page 305
[612] Morehead 2012 (Kindle), Loc 187
[613] Alley 2003 (2018 edition), page 277
[614] Bland 2022, p 367

like lionesses.

It is my suspicion is that if a female *H. sapiens* mates with a male of a different *Homo* species (or a male that is itself a hybrid), the offspring will be subject to these growth-limiting genes and not be subject to the "growth dysplasia" we might initially expect. So a Patrick may not be particularly tall, although feline hybrid models are not likely to apply to the *Homo* genus. It should be noted that in addition to the growth-limiting genes that would contribute to the adult size of the offspring, female humans also carry growth-limiting genes that act specifically upon the fetus within the pregnant mother to protect her health and future fertility. So Patrick's neonatal size would also have been manageable by his tiny mother.

Considering how size is controlled in matings, including hybrid ones, I would not, for the reasons described above, expect the product of a male Sasquatch mating with a female human to be particularly impressive. A 5'-2" adult height for Patrick was probably still a few inches taller than Hahissiat—exactly what we might expect with the action of fetal and adult growth-limiting genes.

However, a male *Homo sapiens* mating with a female of another *Homo* species may produce a hybrid bigger than either parent, like the product of a female tiger and a male lion. A research avenue to test this hypothesis would look at the very tallest of Sasquatch sightings. Do they tend in appearance towards looking very human? Are these 90th percentile Sasquatch more light-framed than the shorter, but more robust adult males? Do they have less-thick hair than the shorter males?

What about Zana, if indeed she was a Eurasian version of Sasquatch? She is thought to have experienced multiple matings with human males. However, her son, Khwit, was not particularly tall. His appearance does not indicate a mating released from growth-limiting genes. I am not able to make a confident prediction of size outcomes from a male *Homo sapiens* mating with a female of another *Homo* species or *Homo* hybrid. There are likely many potential genetic outcomes, particularly if we are dealing with heavily-hybridized populations.

On the other hand, the alleged male product of a Chinese

woman and a Yeren (pictured on page 200) was indeed almost abnormally tall for that time in China (~6'-5" or 2.0 m). But he also appeared very skinny, so he possibly did not weigh much more than a normal adult *H. sapiens* male.

In the context of this book and its core theme of the Patrick story, I have not been able to definitively place Sasquatch in the evolutionary story of hominids. However, this research has turned up interesting leads that point to some origin routes for Sasquatch that appear evolutionarily plausible. I will explore those origin routes further in Volume II.

CHAPTER TWENTY-THREE

The Aging Patrick

We diverged from Patrick's life story on October 10th, 1928 when he was arrested, leaving behind precious mug shots. I believe he was 35 years old at that arrest—almost precisely halfway through a reasonably long life span for a man of that time. Would he now take better care of himself, focus on the productive farm he was so good at managing, be a respectable father to his growing family, and a caring husband to his loyal wife?

Let's find out.

Not too long after Patrick's arrest, Catherine, Pierre Paul's mother, apparently died of influenza on February 15th, 1929 in Loomis, Washington—about 50 miles north of the reservation. She had been living alone for some reason at the recorded age of 77 (though she may have been older).[615] Perhaps she no longer found her son's house a suitable place to live.

A stock market crash came in the fall of that year, October 1929. Stocks on Wall Street probably had little bearing on Patrick's life, but the following deep economic malaise of many years undoubtedly dropped the price for the grain and cattle sales that his family depended on.

Soon in the next year some good news: The birth of Madeline

---

[615] 1929 Indian Census; Colville and Spokane Reservations; Deaths

Christine to Louise and Patrick on March 23, 1930, their fifth child and third girl[616]

Ed Fusch's informant, Louie, reported that he worked for Patrick on his farm from about 1925 to 1930.[617] I suspect that the farm ceased to be something Patrick put much energy into towards the end of that time and that's why Louie's employment stopped. Perhaps the general hard times that began with the stock market crash, and the following Great Depression, forced a reduction in the farm's overhead.

The 1930 Federal Census shows Patrick as head of a household in Nespelem, Washington that held his father (actually *stepfather*) Pierre Paul, wife Louise, and the four surviving children: Ernst, Marie, Stella Lucy, and the month-old Madeline Christine. Patrick is recorded as able to read and write; an English speaker self-employed in the industry of "general farm;" and living in his own home then valued at two hundred dollars.[618]

In approximately May of 1931, hard on the heels of the birth of Madeline Christine in March 1930, Louise gave birth to Isaac H.

Patrick and Louise now had five living children.

As an example of how unreliable the documented ages were for older Native Americans on reservations at this time, Pierre Paul, recorded as 87 years old in the 1930 Federal Census was listed as 62 years old in the April 1932 Indian Census with an estimated birth date of 1870. It was much easier to trace and document this family through names than through years of age.

That same 1932 Indian Census also showed Pierre Paul's stepdaughters, Alice Marie and Madeline, age 20 and 17, continuing to live with him although their mother, Elizabeth, had died back in 1924.

Oddly, the April 1932 Indian Census did not list the last born, Isaac H., at all.[619]

About this time, something was beginning to change for Patrick and he began a transition away from the accomplished and

---

616 1929 Indian Census; Colville and Spokane Reservations; Births

617 Fusch 1992a, page 22

618 1930 United States Federal Census

619 1932 Indian Census; Colville and Spokane Reservations

successful farmer he had been, although as late as the Fall harvest in September 1932, Patrick won a prize for grain at the Nespelem 4-H Club Fair.[620]

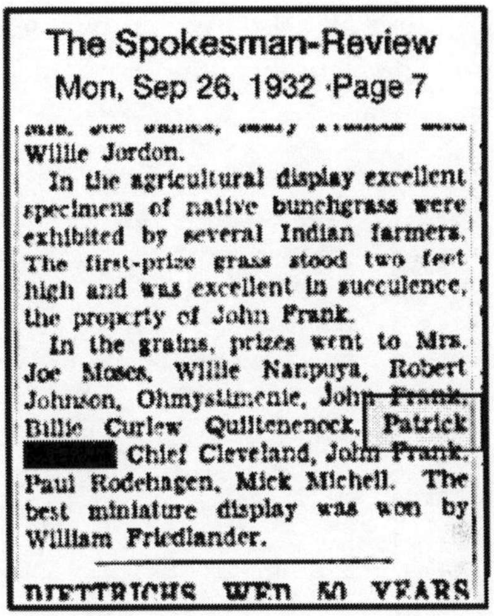
The Spokesman-Review
Mon, Sep 26, 1932 ·Page 7

Willie Jordon.

In the agricultural display excellent specimens of native bunchgrass were exhibited by several Indian farmers. The first-prize grass stood two feet high and was excellent in succulence, the property of John Frank.

In the grains, prizes went to Mrs. Joe Moses, Willie Nanpuya, Robert Johnson, Ohmystimenie, John Frank, Billie Curlew Quiltenenock, Patrick [redacted] Chief Cleveland, John Frank, Paul Rodehagen, Mick Michell. The best miniature display was won by William Friedlander.

DIETTRICHS WED 50 YEARS

In the only record of son Isaac H that I have been able to find—the Internment Records of the Catholic Diocese of Spokane—Isaac H. of Nespelem Washington, with parents Partrick and Louise, died on November 19th, 1932. He was buried on the same day.[621]

The April 1933 Indian Census shows Louise and Patrick with their four surviving children, but Pierre Paul's stepdaughters are not listed.[622] I have not located either of the two young women after 1932.

On November 23rd, 1933, Patrick's stepfather, Pierre Paul, died.[623] Cause of death on his Death Certificate was entered as "Believed to be a paryletic [sic] stroke – No Dr. was in attendance".[624] In a later Supplemental Indian Census Roll the cause of death is listed as cardiac failure.[625] This disagreement in cause of death implies that he might have been found collapsed, perhaps already dead, and no one knew for certain how he died. In the former document his age is given as 73; in the later 63. I think that 63 years old might have been closer to the truth, based on Hahissiat/Madeline's age at abduction. Pierre Paul's birthplace is listed as the Colville Reservation. It is from the Death

---

620 *The Spokesman-Review*, September 26, 1932, page 7
621 11/19/1932, Isaac H. [redacted]; Record of Internments; Catholic Diocese of Spokane
622 1933 Indian Census; Colville and Spokane Reservations
623 April 1, 1934 Indian Census Roll; Colville Agency-handwritten note
624 11/23/1933, Pierre Paul [redacted]; Certificate of Death; Bureau of Vital Statistics; Washington State Board of Health
625 1/01/1938 Supplemental Census Roll; Colville Agency

Certificate that I found Pierre Paul's father's name as "*Antline* or *Squin-i-met-sa*" and his mother's maiden name as "*Nan-u-qui-ya*", eventually just known as Catherine. Relatives and friends provided undertaker care and he was buried the next day, on November 24, 1933.

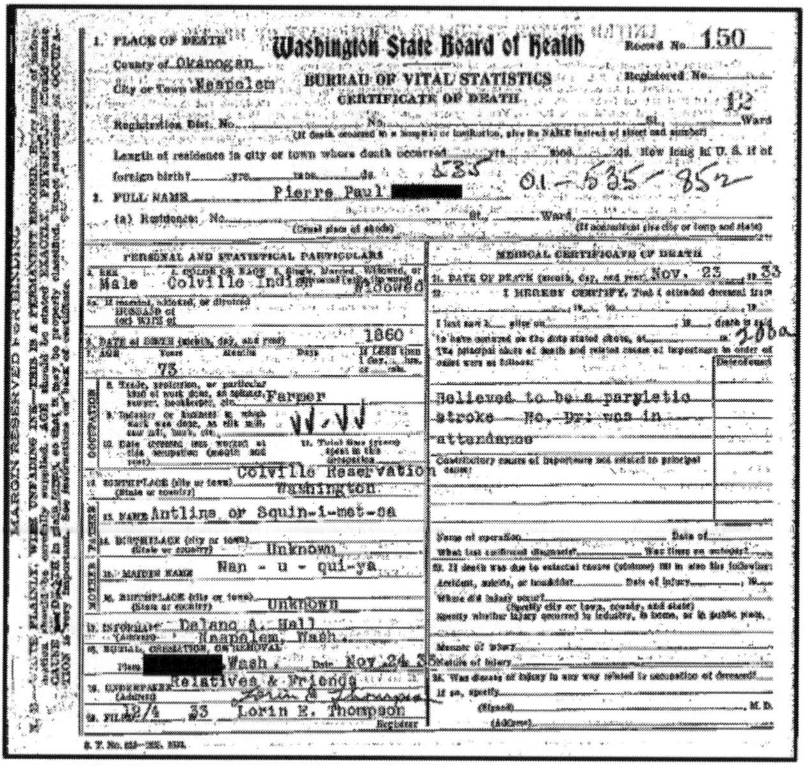

Washington State Board of Health
Bureau of Vital Statistics
Certificate of Death

1. Place of Death: County of Okanogan; City or Town Nespelem — Record No. 150 — 12

2. Full Name: Pierre Paul [redacted]

535 01-535-852

Personal and Statistical Particulars: Sex Male; Colville Indian; Widowed

Date of Birth: 1860 — Age: 73

Trade, profession, or particular kind of work done: Farmer

Birthplace: Colville Reservation, Washington

Father — Name: Antline or Squin-i-met-sa; Birthplace: Unknown

Mother — Maiden Name: Nan - u - qui-ya; Birthplace: Unknown

Informant: Delano A. Hall, Nespelem, Wash.

Burial, Cremation, or Removal: Place [redacted] Wash. — Date Nov 24 33

Undertaker: Relatives & Friends

Filed 12/4 33 — Lorin E. Thompson, Registrar

Medical Certificate of Death: Date of death Nov. 23, 1933. Believed to be a paralytic stroke. No Dr. was in attendance.

Certificate of Death for Pierre Paul, Patrick's step-father. Died November 23, 1933.

Patrick was now the eldest male of the immediate family, with a wife and four children; with all of the responsibilities that role implies.

The Patrick trail peters out until 1935 when we find the town of Nespelem looking less like a Wild West frontier stop and more like a small rural town with a modest newspaper. This announcement was in the January 16th, 1935 issue of the *Nespelem Tribune*:

> A daughter was born to Mr. and Mrs. Patrick [redacted] Saturday, January 12. They have named her Rosalia.[626]

Rosalia would also be known at various times as Mary, Rose, and Rosalie. She was the seventh known child of Patrick and Louise.

Nespelem, Washington about 1935

On March 16, 1935, Nespelem residents voted 49 to 7 to incorporate their town. Since Native Americans had finally gained legal citizenship and the right to vote in the 1924 Indian Citizenship Act, perhaps Patrick himself voted in this incorporation election.

On May 1, 1935 our Sasquatch/*Homo sapiens* hybrid again made the local paper (twice!) for events in late April. Under the subheading, "County Sheriff Here":

> Boyd Hilderbrand, county sheriff, was a Nespelem visitor Wednesday. He took Patrick [redacted] to Okanogan, and lodged him in the county jail on charges of being drunk and disorderly.[627]

And under the sub heading, "Stabber is Sentenced":

[626] 1/16/1935, *The Nespelem Tribune*, page 3

[627] 5/01/1935, *The Nespelem Tribune*, page 3

> Arrested a week ago at Nespelem, Patrick [redacted] pleaded guilty to a third degree assault charge yesterday when arraigned in superior court before Judge William C. Brown.
>
> He was sentenced to six months in the county jail, all but the first two months of the sentence suspended. [redacted] was alleged to have stabbed Frank [redacted], inflicting painful injuries.[628]

Unfortunately, Patrick's free publicity made it as far as a Spokane newspaper under the sub heading, "Nespelem Man Jailed.":

> Patrick [redacted], Nespelem, drew six months in the county jail when he pleaded guilty in superior court of stabbing another man several days ago. Four months of the sentence were suspended.[629]

About all that I can write in defense of Patrick's actions that late April day is that he was not picking on a weaker man. Frank [redacted] was born of Native American parents on February 3, 1914 (so 21 years old at the time of the incident).[630] By my reckoning Patrick was 43 in April of 1935. Patrick was twice Frank's age! And according to Frank [redacted]'s October 16, 1942 World War II draft card, Frank was 6'-2" tall (1.88 m) and 180 pounds (82 kg) compared to Patrick's 5'-2" (1.57 m) and approximately 130 pounds (60 kg). A drunk and disorderly Patrick had attacked a young man half his age, 50 pounds heavier, and a foot taller.

An interesting note: Under "obvious physical characteristics" Frank's draft card notes: "Scar on left bycep [sic]".[631] Is it possible that Patrick inflicted the wound that made that scar still etched in Frank's flesh seven years later?

---

628 5/01/1935, *The Nespelem Tribune*, page 1

629 5/02/1935, *The Spokesman-Review*, page 9

630 2/03/1914, Frank [redacted]; Delayed Birth Registration (4/19/1942); Bureau of Vital Statistics; Washington State Board of Health

631 10/16/1942, Frank [redacted]; Registration Card; World War II

A knife scar on the left side of a victim implies right-handedness in the attacker (Patrick?). In more slim support of my suspicion that Patrick was right-handed is the way Patrick dotted his “i”s in his signature (on this page and on the book cover). It appears he would start the dot with an initially heavy hit top left, followed by a dimishing drag down and to the right. That motion is natural for a right-handed person.

Frank [redacted] went on to live a life of 88 years, not dying until 2002. How many of his children, five grandchildren, and five greatgrandchildren got to hear the story of his knife fight with Patrick, the man rumored in Nespelem to have had a Skanicum father?

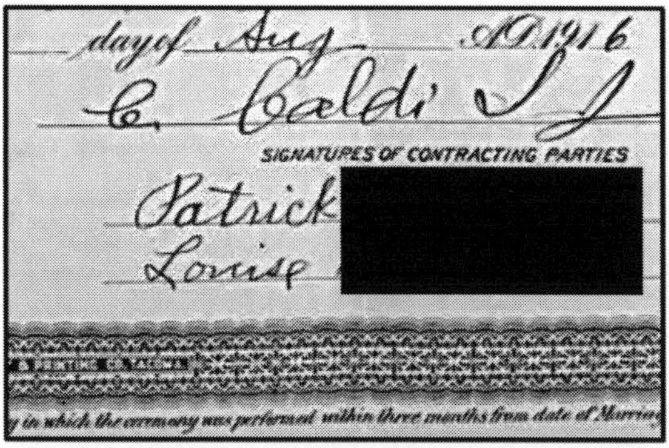
day of Aug A.D. 1916
C. Caldi S J
SIGNATURES OF CONTRACTING PARTIES
Patrick [redacted]
Louise [redacted]
in which the ceremony was performed within three months from date of Marriag

Patrick's signature on his 8/24/1916 marriage certificate to Louise.

Unfortunately, the next trace that Patrick leaves us in the records is with a 5:00 PM booking into the Yakima, Washington county jail on September 27, 1936. He is recorded as a 44-year-old Indian laborer from Nespelem with black hair and eyes, 5’-0” tall, and 130 pounds. He was released two days later, but only after posting an astounding—for the time—$500 bail.[632] Remember: His family home in the 1930 Federal Census was assigned a value of only $200. I am just speculating, but Patrick’s problems with the law may have cost his family the entire ranch.

The charge for this 1936 booking? “Liquor to Indians.”

---

[632] 9/27/1936, Patrick [redacted]; Jail Records; Yakima County; Washington State, U.S., State Corrections and Jail Records

To give Patrick some benefit of the doubt, he may have just sold one bottle to a friend. Or perhaps it was a lucrative side business venture rationalized as defiance of the white invaders and their laws. We don't know. What we do know is that alcohol has been extremely destructive to individuals and of First Nations' culture as a whole since Europeans hit the North American shores. Liquor was a terrible problem for Patrick personally.

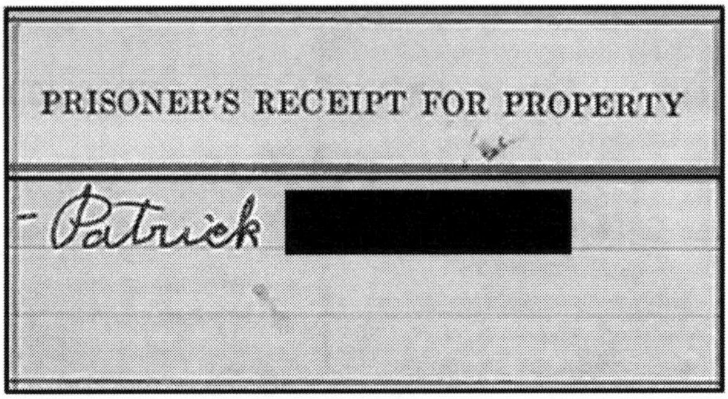
PRISONER'S RECEIPT FOR PROPERTY

-Patrick [redacted]

Patrick's signature from his 9/29/1936 release from Yakima, Washington county jail. Compare this signature for similarity with his marriage certificate signature above, over twenty years earlier.

From the September 30, 1936 *The Spokesman-Review*:

> YAKIMA, Sept 29.—Patrick [redacted], Indian, arraigned before Federal Court Commissioner Thomas Granger here today, pleaded guilty of disposing of whiskey to Dave Condon, Indian, and was released in $500 bond.[633]

Patrick did not turn his life around after the cost and embarrassment of his jailing in Yakima and the newspaper publicity around the circumstances. Less than a year later he was back in jail by midday on March 24th, 1937, again for "Liquor to Indians." I believe that by now Patrick was bringing pint liquor to the Yakima hops yards and selling it at profit to other Indians working there as laborers. He is still recorded as 130 pounds (59 kg), but has grown to 5'-4" (1.63 m). I suspect they took his word on his height during this booking. This time he was waged a fine

[633] 9/30/1936 *The Spokesman-Review*, page 28

of only $100, but also given a 60-day sentence.

On March 25, 1937 Patrick's name was again in *The Spokesman-Review* with "Eight others, who pleaded guilty of supplying liquor to Indians" from the same incident above.[634]He had served less than a month of the 60-day sentence before being discharged on 4/15/1937 and "Taken to Hospital By Dr. Nagler."[635] I don't know why Patrick had to go to the hospital in this instance, but later in life he did suffer from a serious chronic illness.

May 4th, 1937 and Patrick is again booked into the same Yakima county jail. It is only 10:00 AM, but records show that he was first taken into custody at the St. Elizabeth Hospital, which often housed infectious diseases. I suspect that he had been at this hospital since Dr. Nagler had cut short his jail time on April 15th, 1937. Unfortunately, he made the Spokane paper a second time on 5/05/1937 with this re-booking for the same 3/24/1937 offense.[636] He was finally released from jail 6/22/1937, upon "Expiration of Sentence".[637]

Patrick's escapades take a break for the summer. Perhaps he still had the farm back in Nespelem and was busy getting the crops in and harvested, but by Fall he was in trouble with the law again. On November 10th, 1937, when Patrick was 45 years old by my calculations *and* by the arrest record, he was again booked, but this time in Okanogan county (of Washington state). This time he is 5'-1". Refreshingly, he is apparently booked with a friend or co-worker, Tom Williams, also an Indian from Roseburg, Oregon. The offense for both was trespassing. They were released to Evan Jones (perhaps an employer?) the next day. [638] The records show that they had two meals during their stay behind bars. They probably had to pay for those meals: Charging for meals was the

---

[634] 3/25/1937 *The Spokesman-Review*, page 30

[635] 3/24/1937, Patrick [redacted]; Jail Records; Yakima County; Washington State, U.S., State Corrections and Jail Records

[636] 5/05/1937 *The Spokesman-Review*, page 32

[637] 5/04/1937 Jail Records; Yakima County; Washington State, U.S., State Corrections and Jail Records

[638] 11/10/1937, Patrick [redacted]; Register of Prisoners Confined; Okanogan County; Washington State, U.S., State Corrections and Jail Records

jail custom of the time.[639]

On February 6th, 1938, with Patrick still 45 years old and his wife Louise approximately 42, their eighth and apparently final child was born—Gabriel Joseph. This information was obtained from a death certificate. Other children of Patrick still alive, and presumably all still living in the same home, were Ernst 14, Marie 12, Stella Lucy 10, and Madeline Christine 7. These five children all had ¼ Skanicum genetics. Any children of their own in the future would have their Skanicum genetics diluted to 1/8th.

Writing of children, it probably did not help the overall well-being of the family when, on the afternoon of September 5th, 1938—just seven months after the birth of their last child, Gabriel—*both* Patrick and Louise were arrested in Yakima county as part of a group of four. The two men were each identified as "hop picker" and the women as "domestic". The charge for each in cursive appears to read, "Invest." Possibly public drinking. All four were released the next day.[640]

Perhaps this was a bonding experience because the next time we hear of Patrick, less than six months later, it is again with his wife Louise in the February 20, 1939 *Spokane Chronicle*: "Patrick [redacted], 48, and Louise [redacted], 42, both of Nespelem, were held in Grand Coulee jail for federal officers on charges of possession of liquor on Indian land."[641]

A year later, the marital bliss had lost its shine. This from the 2/07/1940 *The Spokesman-Review*:

> **Toppenish, Wash., Feb. 6.**—Patrick [redacted], Toppenish, is serving out a 30 days' sentence in the Yakima County jail for allegedly assaulting his wife and his nephew with a butcher knife. Police said neither was seriously injured. [redacted] was sentenced in justice court by O. B. Root yesterday.

---

[639] 11/11/1937, Patrick [redacted]; Prisoner's Board Record; Okanogan County; Washington State, U.S., State Corrections and Jail Records

[640] 9/05/1938, Patrick [redacted] & Louise [redacted]; Record of Prisoners Committed to County Jail; Yakima County; Washington State, U.S., State Corrections and Jail Records

[641] 2/20/1939 *Spokane Chronicle*, page 15

In this article, Patrick's last name given is one of the two less-used spellings. All three last name spellings differ only by changes in the same single letter. These alternate spellings confused me at first until I noted the surrounding consistency of family structure, date, location, and biography. Toppenish is less than twenty miles south of Yakima, Washington—a location we have placed Patrick in more than once already. At that time, hops and other area agricultural products created a heavy seasonal demand for labor.

I do not know who the nephew is in this account. I never saw mention of his half-sister, Christine, or his stepsisters, Alice Marie and Madeline, as ever having any children. I suspect that this nephew was the child of one of Louise's siblings. I suppose that we may be thankful that this nephew was present—he possibly diffused and diluted Patrick's anger over whatever sparked this attack.

Although I found no evidence that Louise ever officially divorced Patrick, we will find later that she will soon begin to use a different last name. Patrick, at 47 years old—with his drinking and his violence—was fast becoming the scandal of the family and perhaps even of Nespelem.

It was also about this time (according documentation I will cite later) that Patrick and Louise's daughter, Marie (sometimes Masie), moved to Seattle at the age of 15. Was this move due to tension in the home between the parents?

In a turn of events, Patrick made some sort of Social Security application the month after attacking his wife and nephew—in March 1940. In 1940, the U.S. retirement age was 65. Social Security numbers had been given out as early as November 1936. Monthly Social Security (SS) benefits for retired persons had just begun in January of 1940. Patrick was (by my reckoning) 47 at the time of this March application. But his birthdate is entered as September 4, 1888 in this record, claiming an inflated age of 51.

I like to think that Patrick was trying to do better by his wife and family in making this application. He had made many mistakes so far, but I believe he was actually quite shrewd and perhaps was trying then to turn things around for himself and his wife and children.

The few other facts on this SS application confirm to my satisfaction that this was our hybrid. Birthplace: Chelan, Washington. Father: Pierre P. Mother: Madeline Hiasa [I do not know if this last name is simply due to a keystroke limit in this record or an odd blending of Patrick's mother's Anglo and Indian first names].

This application gives Patrick's Social Security number!

Think about that for a moment... While we argue endlessly over whether Sasquatch is an elusive mountain ape or a shapeshifting mind-reading alien, a half Skanicum/half Colville Indian was granted citizenship in 1924 with the Indian Citizenship Act. Then our hybrid was issued an SS card in 1940—over 85 years ago at the time of this writing in 2025, and 27 years before Patty was filmed on Bluff Creek! Pardon my language, but I find these facts *extraordinary*.

He was half Sasquatch!!

The case law precedent implications alone are fascinating.

How is this for extraordinary: In the same month of March 1940, Patrick and Louise are *again* booked together into the same Yakima County jail. It was 9:45 AM on the 29th under the familiar charge of "Invest." Place of residence: Toppenish. Patrick is described as 5'-2" and 140 pounds. Man and wife were both released on the same day.[642]

It troubles me to write that Patrick was *again* jailed at 6:00 PM on the same day! Same charge, but alone this time. He was held overnight and entered in the log book as a 50-year-old laborer.[643] I think he was actually only 47, but by this time Patrick had been given and gave so many birthdates that he probably had lost track of his own *fabricated* age.

In May of the same eventful year (1940), Patrick and Louise were counted in the United States Federal Census out of Yakima. No children were listed. Presumably the kids were all back on the reservation with relatives while the parents were trying to earn

---

[642] 3/29/1940, Patrick [redacted] & Louise [redacted]; Record of Prisoners Committed to County Jail; Yakima County; Washington State, U.S., State Corrections and Jail Records

[643] 3/29/1940, Patrick [redacted]; Record of Prisoners Committed to County Jail; Yakima County; Washington State, U.S., State Corrections and Jail Records

cash income. Patrick is recorded as 51, Louise 46. In a final confirmation of Patrick's minimal education, this census record has 2 as the last school grade attended. As you might remember, the Nespelem school was struggling with fires, closures, and low attendance when Patrick was a child. Louise apparently made it to the 7th grade. Patrick had worked 39 weeks in the past year and earned $260.[644] You can see now why that bail bond of $500 in 1936 would have been devastating to the finances of his family.

On the 24th of August, 1940, Patrick is back in Yakima jail again under the old charge of "Invest". This time he is without Louise. He is discharged on the same day.

In a seeming digression, I am going to quote a May 1941 article in the Washington state *Kitsap Sun* under the headline, *Searching for Boy*:

> **OKANOGAN, Wash., May 26.—(UP)**—A county-wide search was being conducted today for 11-year-old Delbert [redacted] of Oroville who has been missing since yesterday afternoon. Officers believe he may have hitch-hiked to Okanogan for a swim and was drowned.[645]

In thirteen years, this missing boy would enter the Patrick story alive and well and make a lasting impression. The police suspicion in 1941 that he left Oroville—a place with much water—to go swimming in far off Okanogan—with not as much water—doesn't make much sense.

I wish that I now had more positive news to report regarding Patrick and his family. But not yet. Patrick's rate of incarceration was dropping—and he no longer seemed to get much in the way of punitive fines: I think he had dropped moonlighting as a liquor salesman. But this next report is of Patrick and his 18-year-old son ("Ernie" in the record) being jailed *together* in Yakima County for "Investigation" on the evening of August 20, 1941. They are both recorded as laborers from Nespelem. Patrick is still down as the same 50 years old that he was at the same jail in March 1940.

---

[644] 1940 United States Federal Census

[645] 5/26/1941 *Spokane Chronicle*, page 1

Ernst began his first incarceration at 18 years old. They were both released the next day.[646]

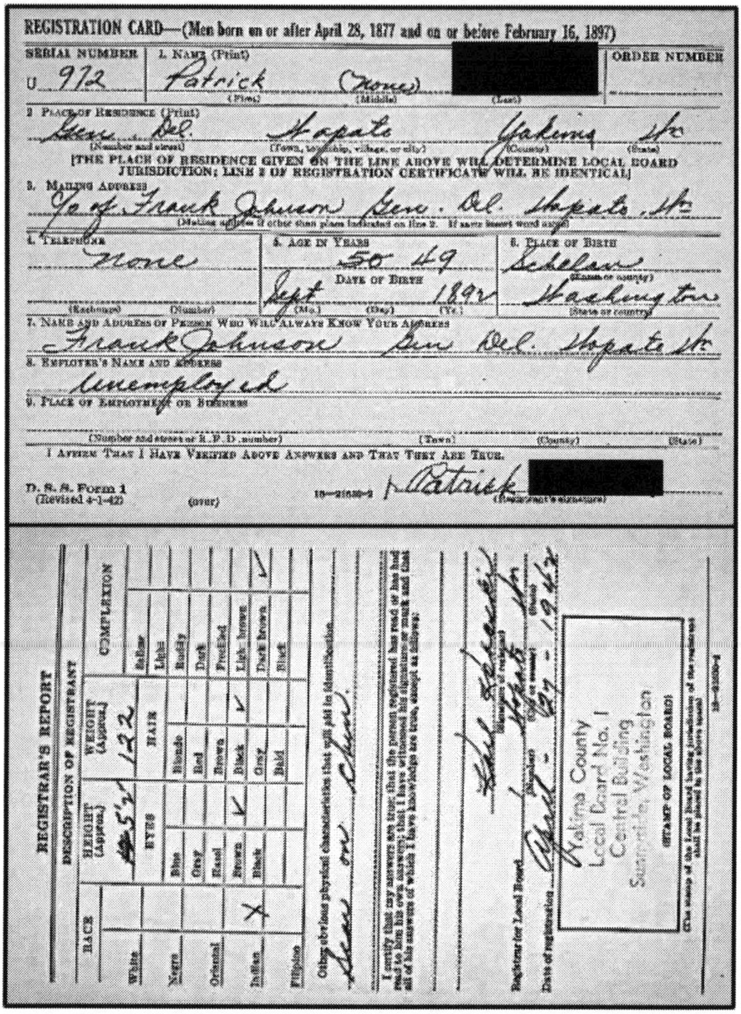
REGISTRATION CARD—(Men born on or after April 28, 1877 and on or before February 16, 1897)

SERIAL NUMBER U 972 | 1. NAME (Print) Patrick (None) [redacted] (First) (Middle) (Last) | ORDER NUMBER

2. PLACE OF RESIDENCE (Print) Gen. Del. Wapato Yakima Wn (Number and street) (Town, township, village, or city) (County) (State)

[THE PLACE OF RESIDENCE GIVEN ON THE LINE ABOVE WILL DETERMINE LOCAL BOARD JURISDICTION; LINE 2 OF REGISTRATION CERTIFICATE WILL BE IDENTICAL]

3. MAILING ADDRESS C/o of Frank Johnson Gen. Del. Wapato, Wn (Mailing address if other than place indicated on line 2. If same insert word same)

4. TELEPHONE None (Exchange) (Number)

5. AGE IN YEARS 50 49 DATE OF BIRTH Sept 1892 (Mo.) (Day) (Yr.)

6. PLACE OF BIRTH Chelan (Town or county) Washington (State or country)

7. NAME AND ADDRESS OF PERSON WHO WILL ALWAYS KNOW YOUR ADDRESS Frank Johnson Gen Del. Wapato Wn

8. EMPLOYER'S NAME AND ADDRESS unemployed

9. PLACE OF EMPLOYMENT OR BUSINESS (Number and street or R.F.D. number) (Town) (County) (State)

I AFFIRM THAT I HAVE VERIFIED ABOVE ANSWERS AND THAT THEY ARE TRUE.

D.S.S. Form 1 (Revised 4-1-42) (over) 16—21630-2 Patrick [redacted] (Registrant's signature)

REGISTRAR'S REPORT

DESCRIPTION OF REGISTRANT

RACE		HEIGHT (Approx.)	WEIGHT (Approx.)	COMPLEXION	
White		5'2"	122	Sallow	
Negro		EYES	HAIR	Light	
Oriental		Blue	Blonde	Ruddy	
Indian	X	Gray	Red	Dark	
Filipino		Hazel	Brown	Freckled	
		Brown ✓	Black ✓	Light brown	
		Black	Gray	Dark brown	✓
			Bald	Black	

Other obvious physical characteristics that will aid in identification: Scar on chin

I certify that my answers are true; that the person registered has read or has had read to him his own answers; that I have witnessed his signature or mark and that all of his answers of which I have knowledge are true, except as follows:

Registrar for Local Board

Date of registration April 27 - 1942

Yakima County Local Board No. 1 Central Building Sunnyside, Washington

(STAMP OF LOCAL BOARD)

Patrick's World War II Draft Card from April 27, 1942. Patrick's height and weight were likely measured that day and may be considered reasonably accurate for him at almost fifty years old.

I lose track of Patrick for over a year before finding a draft card

[646] 8/20/1941, Patrick [redacted] & Ernie [redacted]; Record of Prisoners Committed to County Jail; Yakima County; Washington State, U.S., State Corrections and Jail Records

for *his* second World War, dated April 27, 1942. The Pearl Harbor attack had occurred in December of the previous year, drawing the United States into the war. Patrick's age is first entered as 50, crossed out, and re-recorded as 49 with a birth date of "Sept 1892". No day of birth or middle name was given, as was consistent for Patrick. As you may recall, I believe Patrick was actually born earlier than September—instead about June 1st of the same year. The September birth month claimed was to draw attention away from his mother's abduction as the true origin of her pregnancy. His birthplace: "Schelan" [sic], unemployed, and living in Wapato, Washington. Wapato is between Yakima and Toppenish. Height and weight 5'-2" (1.57 m) and 122 pounds (55 kg) with a scar on his chin. His signature looks much the same as those pictured earlier.[647] I believe numbers for height and weight were likely measured accurately for this draft card and 5'-2" is the adult height I accept for Patrick.

Like father, like ¼ Skanicum son: I also found a draft card for Ernst [redacted] from July of the same year, 1942. He was now living in the big city of Seattle, perhaps by the example of his younger sister, Marie, thought to have moved there herself in 1939. Ernst gives his contact as "Louise Mathews." At first I thought that this Louise might be his mother—finally dropping Patrick's name—but I believe this other Louise was born about 1910 and was approximately 32 years old at the time Ernst registered for the draft. Most likely this Louise was just renting him a room. Ernst's place of employment was the "Western Gearing Works" [actually Western Gear Works: "We Make the Gears That Turn the Wheels of Industry"] and he was a relatively tall 5'-6" (1.68 m) and a slim 137 pounds (62 kg).[648]

Oddly, Patrick makes a second Social Security application or claim in October of 1942. Perhaps this was a disability claim. This time his birthdate is recorded as "1 Sep 1892"... close to accurate. Birthplace is Chelan, Washington. Father and mother's names "Paul" and "Madelein Nowkopop". This document lists a different SS number from the March 1940 document over two

---

[647] 4/27/1942, Patrick (none) [redacted]; Draft Registration Card

[648] 7/02/1942, Ernest (none) [redacted]; Draft Registration Card

years earlier.

Patrick's life since his first arrest in 1928 was marked by repeated jail times and the implied slow disintegration of his nuclear family. I presume the farm was lost, but strong ties remained with Nespelem, even while the family members scattered to other parts of Washington state and experienced an urban lifestyle. The next chapter, beginning in 1943, will begin with another disquieting family event that will also give us some insight into how Patrick's story continues even until today.

CHAPTER TWENTY-FOUR

Patrick's Death

Patrick had at least eight children through his marriage with Louise. Each of those children were ¼ Skanicum blood. Did they show any peculiar traits? According to Ed Fusch's Nespelem informants in the mid-1980s, the two known surviving daughters then had wide mouths, protruding teeth, and "squint eyes." Madeline [Christine], thought to live on the Washington coast, was described by Ed as having "very distinct Skanicum features such as sloping forehead, long peaked ears, etc."[649] It would be helpful to have pictures of some of Patrick's children.

As a matter of fact...

On November 7, 1943, *two* of Patrick's children—Ernst 20 and Marie 19—were arrested in Seattle. Ernst was charged with Disorderly Conduct and Marie with being Drunk and Disorderly.[650] There was a third person involved—prisoner #26000 in the Seattle Police Department system. #26000 was a man originally from Kansas, but with an arrest record in Chicago; Oakland, California; and three previous arrests in Seattle. Three

---

[649] Fusch 1992a, page 21

[650] 11/7/1943, Ernest [redacted] & Marie [redacted]; Mug Shots & Criminal Record; City of Seattle; Washington State, U.S., State Corrections and Jail Records

years before he had been arrested in Seattle with another man and identified as "susp drunk rollers".

My guess is that Marie was under the influence that November day and the third man tried in some way to take advantage of Marie. Ernst was close by and came to the defense of his little sister. General mayhem ensued. Ernst and Marie both have the same note on their Criminal Record sheets: "Cutting scrape with 26000..." 26000's entry was in cursive and a bit hard for me to decipher, but it looks like: "11-7-43 Sea – Inst [?] cutting – Rel." **Sea** is Seattle, of course. I think the final **Rel** implies that 26000 was quickly released. Ernst was held for two days, sentenced to 10 days, all apparently suspended. Marie was initially held for three days before being sentenced to ten days total. It seems Marie had to serve out her jail time. Apparently the 4'-10", 118-pound, ¼ Skanicum woman was the most dangerous of the lot. #26000 picked the wrong woman to mess with.

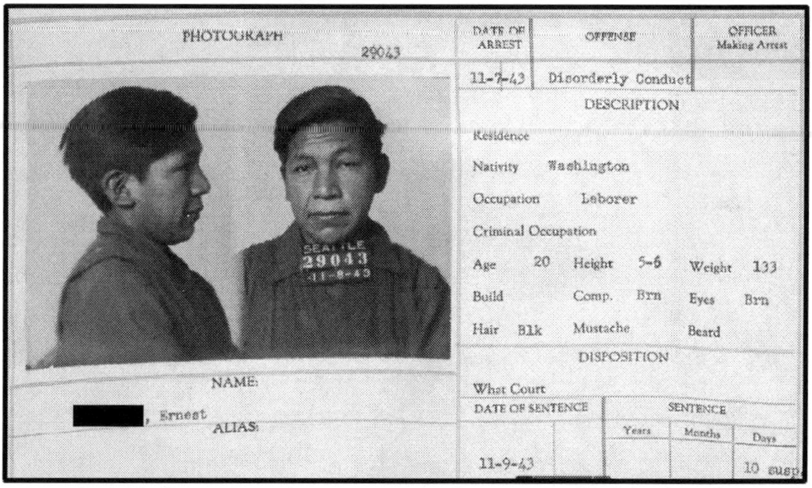

Ernst's 11/7/1943 Seattle arrest record. Age 20; 5'-6" (1.65 m); 133 pounds (60 kg). Ernst's record included no notes of future arrests.

These mugshots give us some ability to draw conclusions on appearance. Ernst is more "normal" looking than his father. His ears have a closer to normal attachment point—higher on his head than Patrick's ears. His forehead does not slope backwards as steeply and his skull is perhaps "normally" high, but the anterior brain case does seem to extend further back than the *H. sapiens* norm, just like Patrick's did. There is some asymmetry in Ernst's face with his left eye and ear placed high relative to those on the right side. But this type of facial asymmetry is common, just politely overlooked.

His ears do not seem all that long or peaked. He certainly does not seem to have an excess of facial hair relative to other Native Americans. The ears may be rotated backwards slightly more than the norm, but not to the extent of Patrick's 41° rotation from the vertical. Ernst's eyes, though—particularly his left eye—are a bit on the squinty side, as Fusch's informants claimed. The oddest feature I find on Ernst is the wrinkles in his forehead. Normally, to create those wrinkles we also open our eyes wider and lift our brows. But Ernst has the wrinkles without opening his eyes wide. Forehead wrinkles are sometimes reported on Sasquatch.

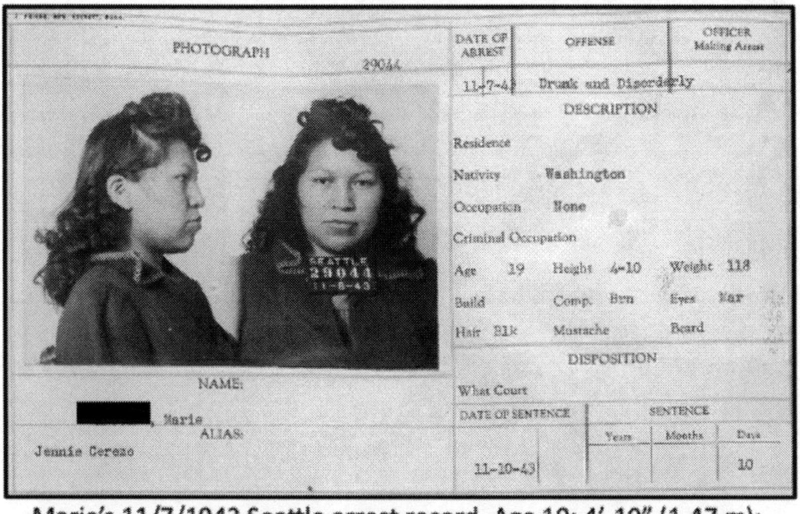
PHOTOGRAPH 29044
DATE OF ARREST: 11-7-43
OFFENSE: Drunk and Disorderly
OFFICER Making Arrest:
DESCRIPTION
Residence
Nativity Washington
Occupation None
Criminal Occupation
Age 19 Height 4-10 Weight 118
Build Comp. Brn Eyes Mar
Hair Blk Mustache Beard
DISPOSITION
What Court
DATE OF SENTENCE: 11-10-43
SENTENCE: Years Months Days 10
NAME: [redacted], Marie
ALIAS: Jennie Cerezo

Marie's 11/7/1943 Seattle arrest record. Age 19; 4'-10" (1.47 m); 118 pounds (54 kg). Note the alias, "Jennie Cerezo". I have not been able to find additional records under that name. Marie's record also had notes on four future arrests.

Marie also does not have much in the way of extraordinary features. Her ears are also rotated a bit towards the back, are smallish, and are perhaps set a bit low. It is hard to determine the shape of her skull with all that head hair. Normal hair pattern.

Eyes not particularly squinty and mouth not all that wide.

Two other features that Ernst and Marie have in common are short necks and weak, though not absent, chins.

Their only features that stand out, two generations removed from a Sasquatch father, are short necks, and, in Ernst's case, the wrinkles in his forehead. I would disagree with the assessment in Ed's paper that the children had "very distinct Skanicum features such as sloping forehead, long peaked ears, etc."

I found no trace of Patrick's family in 1944 and it is not until October of 1945 that the family name comes up again with Marie's booking in Seattle on the 17th of the month for "Disorderly Conduct & Theft". She was sentenced to "$30 or 10 days".[651]

Stella Lucy, Patrick and Louise's fourth child, was granted a marriage license to Ray M. [redacted] on November 10th, 1945 in King County, Washington. They would wait almost a year before getting married on October 8th, 1946.[652] Ray was born in the Philippines about 1899. That would make him about 28 years Stella's senior. As of the 1940 Census, Ray was a single Seattle resident, worked then in manufacturing, and was not able to attend school past the 4th grade.[653]

On March 18th, 1947, Marie was booked in Seattle for Vagrancy. For not having a home, she found a home in the Seattle jail for 90 days.[654]

On September 1st, 1947, Marie was again booked into the Seattle jail, this time for being drunk. She was fined $20.[655]

For the last of her stays in the Seattle jail, Marie, third child of

---

[651] 10/17/1945, Marie [redacted]; Addendum to 11/7/1943 Mug Shots & Criminal Record; City of Seattle; Washington State, U.S., State Corrections and Jail Records

[652] 10/08/1946, Stella Lucy [redacted] & Ray M. [redacted]; Marriage Certificate; King County; Washington State

[653] 1940 United States Census, Seattle

[654] 3/18/1947, Marie [redacted]; Addendum to 11/7/1943 Mug Shots & Criminal Record; City of Seattle; Washington State, U.S., State Corrections and Jail Records

[655] 9/01/1947, Marie [redacted]; Addendum to 11/7/1943 Mug Shots & Criminal Record; City of Seattle; Washington State, U.S., State Corrections and Jail Records

Patrick, was booked on September 26th, 1947, for Disorderly Conduct and fined $40.[656]

In the next month, Delbert Dee [redacted] (our missing eleven year old in 1941, found again), begins over eight months in jail in Okanogan County (the county of Nespelem).[657]

A few months later, on February 17, 1948, a marriage license is granted to Melchor K. [redacted] and Marie, the "cutting scrape" daughter of Patrick. The two would not marry until three months later in the Seattle Courthouse on the 29th of May when Marie was 23 years old. Stella, who had married Ray [redacted] on 10/8/1946, less than two years previously, signed as one of two witnesses to Marie's wedding, but under her maiden name.[658]

Perhaps the delay in Marie's wedding had something to do with Melchor Kasko's arrest—four days after the marriage license was granted—for suspected burglary along with five other people (none of those five were his fiancé).[659] Melchor had a Filippino father and an Alaskan Native mother and must have towered over Marie at 5'-10" tall. Marie did not officially divorce Melchor until almost 35 years later, when they were living in different states and Marie finally petitioned him for a divorce. However, at least as early as 1960, the Seattle City Directory indicated that Melchor was living with a different woman, Maura Delacruz. The divorce certificate indicated that there were no children as a product of Marie's marriage to Melchor.[660] Melchor Jr. lived until 2004.

I never found a divorce decree for Stella's marriage, but I suspect, from the use of her maiden name in 1948, that the 1946 marriage to Ray M. did not last long. However short of a marriage,

---

[656] 9/26/1947, Marie [redacted]; Addendum to 11/7/1943 Mug Shots & Criminal Record; City of Seattle; Washington State, U.S., State Corrections and Jail Records

[657] July 1948, Delbert D. [redacted]; Prisoner's Board Record; Okanogan County; Washington State, U.S., State Corrections and Jail Records

[658] 5/29/1948, Marie [redacted] & Melchor K. [redacted] Jr.; Marriage Record; King County; Washington State

[659] 2/21/1948, Melchor K. [redacted] Jr.; Mug Shots, Washington State Corrections and Jail Records

[660] 3/02/1982, Marie [redacted] ([redacted]) & Melchor K. [redacted]; Certificate of Dissolution or Declaration on Invalidity of Marriage; Washington Divorce Records

later evidence suggests that there was a child from her relationship with Ray born in 1946 or 1947. This suspected child was perhaps Patrick's first grandchild when Patrick was still in his mid-fifties. Patrick was only about seven years older than Ray himself. The evidence for this grandchild, a boy, is slim, but I will present it later.

Also in 1948, Madeline Christine appears in the Seattle Telephone Directory, living with William O. [redacted] (a professional driver), in a house in north Seattle, very close to what is now the historic Gas Works Park on Lake Union.[661] I was able to find a WWII Draft Card for William Oliver (with the same last name as William O. above) from six years earlier, registered at the same Seattle address and he is still giving that address on a later Korean War draft card. William Oliver was born in 1923 and was six years older than Madeline Christine.[662] I do not know what type of relationship Madeleine Christine had with housemate William Oliver, but in 1977 documentation, William Oliver is described as never married by his sister. Despite this, Madeline Christine will eventually have his last name recorded as her own. Confusing to me: Despite this documentation of living with William Oliver in 1948, we will soon hear of how Madeline Christine and her *husband*, Joseph, are grieving for their dead infant daughter in early 1949, along with the departed's two *older* brothers, children of the same parents.

Delbert Dee [redacted] is released from Okanogan County jail on July 22nd, 1948.[663]

It is a shame how this information gets to us, but Patrick's movements around the state of Washington during his lifetime can be tracked by where he was jailed (I will eventually provide a map of locations where I know Patrick spent at least one night). This time Patrick appears in the Whatcom County Jail in Bellingham, Washington, north of Seattle. Patrick was booked on 9/13/1948 with a charge of Public Intoxication. He gave an address of 1413 Central Avenue, Bellingham, near the waterfront. He was given a

---

661 1948, Madeline & Wm O [redacted]; Seattle, Washington; City Directory
662 7/30/1942, Melchor Kasko [redacted] Jr., Draft Registration Card
663 7/22/1948, Delbert D. [redacted]; Prisoner's Board Record; Okanogan County; Washington State, U.S., State Corrections and Jail Records

choice of paying a $10.00 fine plus $2.00 in costs or to serve four days in jail. He took the time.[664] [665]

Unfortunately, it seems that as soon as he was released, Patrick was in trouble again. This time we read about the latest in the 9/17/1948 issue of *The Bellingham Herald*:

> Patrick [redacted], 1413 Central avenue, and Leonard Johnnie, Nooksack, were fined $25 and costs Friday afternoon in police court on a charge of disorderly conduct. The two Indians were arrested following a brawl at the marine ways Thursday evening.[666]

I suspect Patrick was released immediately after this brawl. As it turns out, he had family in Bellingham that may have paid his fine and court costs this time or he may have had work income.

Early in the next year (1949) we find another *The Bellingham Herald* article bringing news, good and bad. News of one of Patrick's daughters married; news of *three grandchildren*; confirmation of Patrick's address in Bellingham, and news of a tragic early death.

Servies to Be Held For Janet [redacted]

Graveside services will be held at 10 a. m. Monday in the [redacted] cemetery for Janet Louise, 2-months-old daughter of Mr. and Mrs. Joseph [redacted], 1413 Central avenue, who died Friday in a local hospital following a short illness. Rev. Fr. John McDonnell will officiate, burial under the direction of the Westford-Beck funeral home.

Besides the parents, the child is survived by two brothers, Joseph Jr., and Martin August, at home; a grandmother, Mrs. Magdalene [redacted], city, and two grandparents in Seattle, Mrs. Louise [redacted] and Patrick [redacted].

1/30/1949 funeral service announcement in The Bellingham Herald for Janet, infant daughter of Madeline Christine (Patrick's fifth child) and Joseph ________. Janet was perhaps the *fourth* grandchild of Patrick and Louise. Two previous grandchildren by the same parents are noted in this article: Joseph Jr. and Martin August.

This short article carries a great deal of information about

---

[664] 9/13/1948, Patrick [redacted]; Register of Prisoners; Whatcom County; Washington State, U.S., State Corrections and Jail Records

[665] 9/13/1948, Patrick [redacted]; Confined in the County Jail; Whatcom County; Washington State, U.S., State Corrections and Jail Records

[666] 9/17/1948 *The Bellingham Herald*, page 5

Patrick. It confirms that Patrick's official address in Bellingham is shared with his daughter and her family, including grandchildren in two older boys surviving after the sad death of Janet: Joseph Jr. and Martin August (both 1/8 Skanicum). Oh, what an influence Patrick must have been fon those kids! It is worth noting for further confirmation of identity that Janet had the middle name of Patrick's wife—Louise. I have blacked out last names to protect any living relatives, but Louise is now using a different last name than Patrick.[667] I never found a divorce decree, but Louise might understandably have finally given Patrick his walking papers and Patrick moved from Seattle up to Bellingham to live with one of his more forgiving daughters.

I found a Death Certificate for little Janet Louise. She was only 2 months and 25 days old when she died in St. Luke's Hospital in Bellingham on January 28, 1949 of pneumonia. Her birthdate would then have been November 3, 1948. Janet Louise is identified on the document as "Indian." Her mother's first name on the certificate is spelled, "Madaline" with maiden name an even more poor spelling of Patrick's last name. I am convinced that this grieving mother is Madeline Christine, Patrick's fifth child, born in 1930.[668] According to Ed Fusch's 1992 paper, Madeline Christine—one quarter Skanicum—was still alive in Washington state at least as late as 1985.[669] That information has proven to be incorrect.

From a wedding license, I was able to narrow down grandchild Joseph's date of birth as likely in 1946. That leaves a window until about February 1, 1948 for the other boy's pregnancy and birth. Grandson Martin August was then probably born in 1947.

Patrick was still alive on April 6, 1950, but also still getting into trouble with the law. We find him then on the 1950 Census in the King County Jail as a prisoner and inmate. He is down in the Census as American Indian and 58 years old, though I think he probably had a couple of months before he was actually to turn 58. He is declared as married, though Louise might have wished

---

[667] 1/30/1949 *The Bellingham Herald*, page 17

[668] 1/28/1949, Janet Louise [redacted]; Certificate of Death; Public Health Statistics Section; Washington State Department of Health

[669] Fusch 1992a, page

it wasn't so. Louise was now living independently and staying out of trouble with the law.[670]

The same Census finds a change of fortune for Madeline Christine, now at the Lester Apartments in Seattle. She is listed as the **separated** Head of her household at only 20 years old, with two children, Joseph L (3 years) and Donna M (5 months). Her older brother Ernst (26 years) is also living with them when the census was taken in April. There is no sign or mention of Madeline Christine's husband, Joseph, or of their second (?) child Martin August.[671] Joseph L is presumably the surviving brother, "Joseph Jr" from her marriage with the elder Joseph. Donna M, presumably from the same marriage, was born after the January 1949 death of Janet.

It is possible that Madeline had moved to Seattle to be close to better medical care for her children. Sadly, Donna Marie died on February 10, 1952 in the Tacoma Indian Hospital from tuberculosis meningitis. Donna was perhaps Patrick's fifth grandchild. It looks like Donna inherited her middle name from Madeline Christine's older sister, Marie.

Tuberculosis meningitis is a slow-progressing condition, difficult to diagnose. Donna's initial low-grade fever, headaches, and mood changes might not have been understood or even noticed until they evolved into vomiting and seizures. Donna Marie had suffered from tuberculosis almost her entire life, but the meningitis aspect of her condition may have been overlooked until it was too late to treat. The fact that she was autopsied implies that her caregivers were not sure what had killed her until the autopsy confirmed *Mycobacterium tuberculosis* in the tissues surrounding her brain or spinal cord. The poor little girl had been housed in the hospital for one year and eight months (or at least she was under the hospital's care that entire time). As identity confirmation, her mother's first name is given as "Madeline", with Patrick's last name correctly spelled as Madeline Christine's

[670] 4/06/1950, 1950 United States Federal Census; Washington State; King County

[671] 4/15/1950, 1950 United States Federal Census; Washington State; King County; Seattle

maiden name.[672]

The Tacoma Indian Hospital where Donna Marie (Patrick's granddaughter) was cared for from June 20, 1950 until her death on February 20, 1952.

The two death announcements for Donna Marie mention only being survived by her parents and her maternal grandmother (Louise). There is no mention of her maternal grandfather, Patrick, or her older brother Joseph Jr. who would have been five years old. However, I believe both of these people were still alive when Donna Marie died. I suspect family dynamics might have kept them both out of the announcements.[673]

It seems as if all I report on in this book is death, birth, and imprisonments, but it is mainly records of these events that outlast our brief lives. This next milestone is yet another sad event: The death of Stella Lucy, Patrick's fourth known child on March 11, 1953. According to the Death Certificate, the immediate cause of death was from lobar pneumonia (inflammation and solidification of a lung lobe with accompanying high fever, chest pain, and extreme difficulty breathing). Stella Lucy's chronic pulmonary tuberculosis certainly contributed and weakened Stella's immune

---

672 2/20/195, Donna Marie [redacted]; Certificate of Death; Public Health Statistics Section; Washington State Department of Health

673 2/22/1952, *The News Tribune* [Tacoma, WA], page 22 and 2/25/1952, page 2

system. Tuberculosis was a significant killer of Native Americans into the 1970s and was a particular bane for Patrick's kin. Stella did not die in a hospital or at her residence. Oddly, she was just one-half block off Seattle's historic Pioneer Square. The address of death given was about one mile away from her home address. In 2025, the death location was some sort of literal underground bar. Interestingly, Patrick is given credit as the "Informant" on the Death Certificate. With sadness, the information on the Certificate leads me to suspect that Stella was drinking buddies with her father, while living with chronic tuberculosis, and that she died in a bar from untreated pneumonia.

Stella's age is shown on the form as 26, but the form (or the Informant) got the year of birth wrong (at least as far as older Census information indicates). I believe that she was only 25 years old when she died. The form describes her as single, so apparently her marriage to Ray in 1946 had ended by 1953. I only find one future reference to what I think was a male child of that marriage. Presumably that child, if still living, remained with his father, Ray. Stella is shown as employed "at home" and of the "white" race. Since Patrick was the Informant, this entry implies that Patrick claimed that his daughter was white, although she was actually ¾ Native American and ¼ Skanicum. Her real heritage would suggest that she was probably markedly dark-skinned, and not likely to pass as white, but perhaps I am missing something. The birthplace is given simply as "Washington." Patrick may have realized if he supplied the correct reservation town name of Nespelem, that his fiction regarding Stella being white would look more shaky.

The mother of Stella is confirmed to be "Louise", but the mother does not have Patrick's last name, but instead the alternate name that was first noted in 1949, so I presume the marriage estrangement continued at least into 1953.[674]

In a family plagued by sorrow and death, the death of Stella strikes me as one of the saddest, but this next death is a close runner-up.

---

674 3/11/1953, Stella [redacted]; Certificate of Death; Public Health Statistics Section; Washington State Department of Health

Ernst died on June 29th, 1954 after a fall the previous day from scaffolding while painting. He fractured his skull in the fall. The immediate cause of death was listed as epidural hematoma—bleeding into the cranial cavity. Since he lived for hours after the fall, it is likely that he was initially conscious—with severe head pain—but his speech would have gradually slurred as he experienced increasing dizziness and vomiting. In a fatal case like this it is expected that he eventually lost consciousness and had seizures. His injury might not have been fatal today with CT scans and craniotomies to relieve the pressure on his brain, but even though he was presumably quickly admitted to King County Hospital, he was dead by 4:00 AM the day after the fall.

Ernst was Patrick and Louise's second child, single, apparently childless, and 30 years old.[675]

I have no idea if family was present at his death, although I think Louise and some siblings were all living in Seattle. Probably Patrick was nearby too.

Delbert Dee [redacted] at 24 years old (our missing eleven year old in 1941 and inmate in 1947), married Rosalie (under the frequently used alternate name, Mary Rose), 19, on October 16, 1954. Confusingly, the Marriage Certificate gives Mary Rose's birthplace as "Nespelem, *Montana*".[676]

Gabriel J. [redacted], absent from the records since his birth in February 1936, reappears as he goes on active duty with the army at Fort Ords in California on June 30th, 1955, at the age of 19. He is discharged again, for unknown reasons on September 28th, 1955. This information comes from Gabriel's Korean War Draft Card of April 10th, 1956. I presume that since he served in the Army less than three months, he was not excused from the draft in the next year. His draft card has him at 139 pounds and 5' 4-1/2" tall (1.64 m). Importantly (soon), he gives Mrs. Mary [redacted] as his next of kin (not his father Patrick or his mother Louise). This Mary is his sister Rosalie, sometimes known as Mary Rose, or just Mary; married to Delbert Dee [redacted] for less than two

---

[675] 6/29/1954, Ernst [redacted];Certificate of Death; Public Health Statistics Section; Washington State Department of Health

[676] 10/16/1954 Mary Rose [redacted] & Delbert Dee [redacted]; Marriage Record; King County; Washington State

years now. Gabriel is living on the south side of downtown Seattle. His sister lives about two miles away.[677]

Gabriel Joseph finds himself in trouble before 1956 is over. On 10/05/1956, after arrest by the Seattle Police Department, he is sentenced to 30 days for assault.[678]

There are a few families at this time under the same last name as Delbert's, so I can't swear to the accuracy of this family line, but I believe that Rosalie/Mary Rose/Mary and Delbert Dee went on to have at least four sons, beginning in May of 1957 with the birth of Daryl Dee. You will see some disturbing evidence for the existence of these grandchildren soon enough.

In yet another tragedy for Patrick's clan, Daryl Dee would die just short of three months old from pneumonia on August 22nd, 1957. Daryl Dee's race is entered as "Indian",[679] although his father, Delbert, was white according to the 1940 Census.[680] Delbert was born about 1930 compared to Rosalie in 1935.

The last born of Louise and Patrick's children, Gabriel Joseph, appears again on August 24th, 1957. Note that his infant nephew, Daryl Dee, son of his favorite sister, Rosalie/Mary Rose/Mary, had died just two days earlier. At the age of 19, Gabriel was charged with "Assault in the Second degree" in a downtown Seattle incident involving an unnamed sister and brother-in-law. The sister involved could have been Marie (already known for a drinking problem and frequenting downtown Seattle) or Madeline Christine, both known to be living in Seattle at this time. But I believe the sister in the account that you will soon read is his chosen next of kin, the grieving mother Rosalie/Mary Rose/Mary.

On 12/05/1957 Gabriel Joseph was sentenced to "No More Than 10 years" for Assault in the Second Degree. On 12/10/1957 Gabriel entered the Washington State Reformatory in Monroe, Washington. From the Inmate's Statement (apparently dictated on

---

677 4/10/1956, Gabriel Joseph [redacted]; Registration Card; Selective Service System

678 10/05/1956, Gabriel Joseph [redacted]; Mug Shots; Washington State Corrections and Jail Records

679 8/22/1957, Darryl Dee [redacted]; Certificate of Death; Washington State Department of Health

680 1940 United States Federal Census; Washington State; Okanogan County

or after 12/10/1957), Gabriel describes in his own words the events that led to his imprisonment (there are no edits below from the account as written in 1957, although I offer some bracketed suggestions):

> Aug. 24, 1957, around 6 o'clock I was coming out of the Threater [sic] where I saw a show, and went down town to look for my mother and sister. I met my sister [presumably Rosalie/Mary Rose/Mary] when she just came of the Britannia Tavern where she usually drinks [on Pioneer Square just a 5-minute walk from where Gabriel lived and within a few hundred feet of Stella's death in 1953]. She and my Brother-in-law [Delbert Dee?] were both felling pretty high from drinking. I was going to go home but they insisted on, for me, to have a drink with them. I said no the first time but they would not stop bothering me, [I went] along with them to a tavern were I always went to, and my sister and Brother-in-law bought me (instead of a drink) drinks. I must've stayed there about four hours I guess and I was loaded. We went to a cocktail lounge and bought ourselves a couple of Tom Collins and a bottle of beer each. We, that's my sister and al [sic, Del?], had an argument about me not giving her any money for her sons school expenses. Of course I knew she wouldn't have one. Truth of that amount of money to send her son to a reasonable school. But no she wanted me to pay her for her son's school tuition. Here she has somebody who has a job and money sometimes, and she wouldn't even ask him to help pay for [h]is son's tuition for school. That's how it sta[r]ted to get me mad. I guess I would have done the same to my sister if she was where the man, who I stabbed, was standing. I guess that's what the reason [is] why I stabbed the person. I said I wanted to see if I could make quick money. That's what I said to the Detectives who was questioning me. But that's what came out of my head first, when I spoke out, the reason I stabbed the guy. I guess that's my version of the story, and I don't car[e] what the detectives say. I feel it was a just punishment they gave me. Because I had no right to stab a person who or isn't

in the family.[681]

Gabriel's account, after he had months to come up with mitigating circumstances, boils down to, "I didn't mean to stab that man. I had no right to stab a person who isn't family. I really wanted to stab my sister because she asked me for money." The son they argued over in that cocktail lounge may have been another grandchild of Patrick's born in 1956 to Mary Rose and Delbert Dee. But that child would have been just over a year old at the time of the incident. Quarreling about school tuition for a one-year old, two days after the death of another child seems ridiculous, but, of course, the entire stabbing seems ridiculous. Patrick was known as very intelligent. I am not certain that his last living son, Gabriel Joseph, shared that quality.

Gabriel Joseph (1/4 Skanicum), son of Patrick and Louise, in 1957, 1960, and 1962. Respectively aged 19, 24, and 25. No obvious Sasquatch features.

On April 25th, 1960, Gabriel was transferred to the Larch Corrections Center in southern Washington and on October 24th, 1960 he was released on parole.

In what was perhaps encouraging news for Patrick and Louise,

---

[681] 12/10/1957, Gabriel Joseph [redacted]; Mug Shots; Washington State Corrections and Jail Records

Marie began living with Baptiste Casey [redacted] in 1961. Baptiste was almost 13 years older than the fiery Marie[682] and had been married previously. That first wife would die in 1962, but under a different married name, so it is likely Baptiste and that first wife divorced before Marie and Baptiste began living together. Surprisingly perhaps, Marie would remain with Baptiste until his death. Baptiste had two sons from the previous marriage—stepsons now to Marie.

On September 5th, 1961, Gabriel was back in jail in Pierce County, Washington (the county of Tacoma, Washington) for auto theft, no vehicle license, and driving without a driver's license on his person. He was unemployed when taken into custody, with 16 cents in his wallet. The next day he was released to the Seattle Police Department.[683] I'm not sure why SPD picked Gabriel up from Pierce County Jail, but it was probably not for a complimentary ride home. It may have been to transfer him back into state custody for the presumed parole violation.

I was not able to find additional records of Gabriel until the following year.

On October 21st, 1961, Patrick was admitted or perhaps *committed* to the King County T.B. (tuberculosis) Hospital. In 1961 that hospital was housed in the Firland Sanatorium just north of the Seattle city limits, a decommissioned naval hospital. Known then as "The White Plague", tuberculosis is highly contagious through inhalation of airborne droplets or by swallowing the *Mycobacterium tuberculosis* bacteria. TB spreads easily in crowded conditions and most commonly attacks the lungs.

You may recall that in 1937 Patrick was temporarily discharged from jail in Yakima, Washington by a doctor who transported him to St. Elizabeth Hospital, known for its treatment of infectious diseases (see page 295). Patrick then spent approximately three weeks at that other hospital before being booked back into the Yakima jail to serve the remainder of his

---

682 7/29/1983 Baptiste "Casey" [redacted]; Obituary; *The Missoulian*, page 11

683 9/05/1961, Gabriel Joseph [redacted]; Sheriff's Jail Register; Pierce County; Washington State Corrections and Jail Records

sentence. It is possible Patrick already had tuberculosis in 1937, 24 years before being admitted to Firland.

By the mid-1950s, new drugs had made the successful treatment of TB possible and Firland's mortality rate dropped to just 6%. Patient time spent at Firland was cut in half, but 10% of all patients were alcoholics. As TB became more feasibly and successfully treatable, infected alcoholics began to be involuntarily committed to Firland "without formal legal process." In 1965 that approach was prohibited under court order.

Alcoholics and other "difficult" patients were confined to the infamous Ward Six. I suspect Patrick qualified for housing in this ward both due a lack of cooperation and his alcoholism. Baron Lerner provides some detail regarding Ward Six in his 1998 book, *Contagion and Confinement: Controlling Tuberculosis along the Skid Road*:

> ...located in the old naval brig... it was equipped with both locked doors and heavily screened windows. Included on the ward were seven locked cells, which contained only concrete slabs covered by thin mattresses. Patients admitted to Ward Six (most of whom were intoxicated) spent the first twenty-four hours in one of these cells for the purpose of sobering up or delousing.[684]

By my reckoning, Patrick was 69 years old when he entered Firland Sanitorium for treatment of his tuberculosis. Unfortunately, in a life of missed opportunities, Patrick was not to leave Firland alive.

Just into the New Year, after the beginning of Patrick's likely enforced stay at Firland to treat his tuberculosis, Gabriel Joseph, Patrick's last known child, was again having his picture taken for at least the third time at the Washington State Reformatory (prison) in Monroe. Gabriel was 25 years old. Gabriel would remain imprisoned until after the death of both of his parents, Patrick and Louise.[685]

---

[684] Lerner 1998, page 121

[685] 1/16/1962, Gabriel [redacted]; Mug Shots; Washington State Corrections and Jail Records

Gabriel's parent's hearts literally failed.

Patrick, our remarkable and tragic half-Skanicum miracle, died at 6:00 AM on August 5th, 1962 in Seattle, according to his Death Certificate. Within the timeline that I established of his likely birth around June 1st of 1892, Patrick was 70 years old when he died at the Firland Sanatorium. He did not die of the "moderately advanced" pulmonary (lung-based) tuberculosis noted as contributing to, but not responsible for his actual fatal condition. Death came from cardiac failure due to "many years" of arteriosclerotic heart disease (damaged arteries) that may have manifested in Patrick as chronic fatigue, chest and leg pain, dizziness, and shortness of breath. He lived for approximately one day after the initial cardiac event.

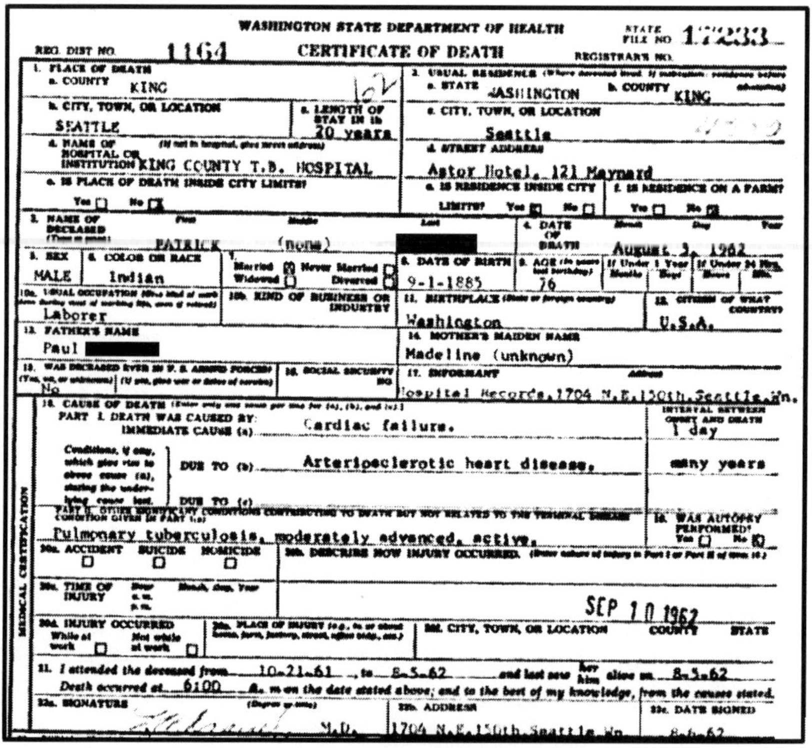
WASHINGTON STATE DEPARTMENT OF HEALTH
CERTIFICATE OF DEATH
REG. DIST NO. 1164
STATE FILE NO. 17233
REGISTRAR'S NO.
1. PLACE OF DEATH a. COUNTY KING
b. CITY, TOWN, OR LOCATION SEATTLE
c. LENGTH OF STAY IN 1b 20 years
d. NAME OF HOSPITAL OR INSTITUTION KING COUNTY T.B. HOSPITAL
e. IS PLACE OF DEATH INSIDE CITY LIMITS? Yes ☐ No ☒
2. USUAL RESIDENCE a. STATE WASHINGTON b. COUNTY KING
c. CITY, TOWN, OR LOCATION Seattle
d. STREET ADDRESS Astor Hotel, 121 Maynard
e. IS RESIDENCE INSIDE CITY LIMITS? Yes ☒ No ☐
f. IS RESIDENCE ON A FARM? Yes ☐ No ☒
3. NAME OF DECEASED First PATRICK Middle (none) Last [redacted]
4. DATE OF DEATH August 5, 1962
5. SEX MALE
6. COLOR OR RACE Indian
7. Married ☒ Never Married ☐ Widowed ☐ Divorced ☐
8. DATE OF BIRTH 9-1-1885
9. AGE (in years last birthday) 76
10a. USUAL OCCUPATION Laborer
10b. KIND OF BUSINESS OR INDUSTRY
11. BIRTHPLACE Washington
12. CITIZEN OF WHAT COUNTRY? U.S.A.
13. FATHER'S NAME Paul [redacted]
14. MOTHER'S MAIDEN NAME Madeline (unknown)
15. WAS DECEASED EVER IN U.S. ARMED FORCES? No
16. SOCIAL SECURITY NO.
17. INFORMANT Hospital Records, 1704 N.E. 150th, Seattle, Wn.
18. CAUSE OF DEATH
PART I. DEATH WAS CAUSED BY: IMMEDIATE CAUSE (a) Cardiac failure. — 1 day
DUE TO (b) Arteriosclerotic heart disease. — many years
DUE TO (c)
PART II. OTHER SIGNIFICANT CONDITIONS CONTRIBUTING TO DEATH BUT NOT RELATED TO THE TERMINAL DISEASE CONDITION GIVEN IN PART I(a) Pulmonary tuberculosis, moderately advanced, active.
19. WAS AUTOPSY PERFORMED? Yes ☐ No ☒
20a. ACCIDENT ☐ SUICIDE ☐ HOMICIDE ☐
20b. DESCRIBE HOW INJURY OCCURRED.
20c. TIME OF INJURY
SEP 10 1962
20d. INJURY OCCURRED While at work ☐ Not while at work ☐
20e. PLACE OF INJURY
20f. CITY, TOWN, OR LOCATION COUNTY STATE
21. I attended the deceased from 10-21-61, to 8-5-62 and last saw him alive on 8-5-62
Death occurred at 6:00 A.m. on the date stated above; and to the best of my knowledge, from the causes stated.
22a. SIGNATURE M.D.
22b. ADDRESS 1704 N.E. 150th, Seattle, Wn.
22c. DATE SIGNED 8-6-62

Patrick's Certificate of Death, August 5, 1962.

Patrick lived a long life compared to other Native Americans of that time—outliving four of his children by Louise, although one of these living children was already hospitalized before

Patrick died and would die unpleasantly just nine days after him.

Incredibly, to me personally, Patrick died in Seattle on the precise day that I entered the United States *and Seattle* as a very young immigrant. On that day I settled with my family into the heavily Scandinavian Ballard district of Seattle just a half dozen miles from where Patrick breathed his last in Firland.

The death certificate in my possession shows his correct last name, lists his mother as "Madeline" (with maiden name unknown), and his father as "Paul" (with the correct last name). Patrick is identified as a married male Indian, so apparently he and Louise never officially divorced despite everything wrong in their marriage.

Things that the Death Certificate did not get correct include reporting that Patrick had lived in Seattle for the previous twenty years. We know that he had a declared address in Bellingham, Washington during this time. It also said that he was born in 1885 (not the likely 1892) and that he was 76 years old (not 70). It also lists his usual occupation as "laborer." I do not think *laborer* does justice to his former happy days as a farmer and rancher back in Nespelem before alcohol took such a hold on him and devastated his family.

Three days after his death, his body was removed and transported to Okanogan, Washington for burial arrangements. Okanogan city is on the edge of the Colville Reservation. After half a lifetime of pain and wandering, Patrick was going home.

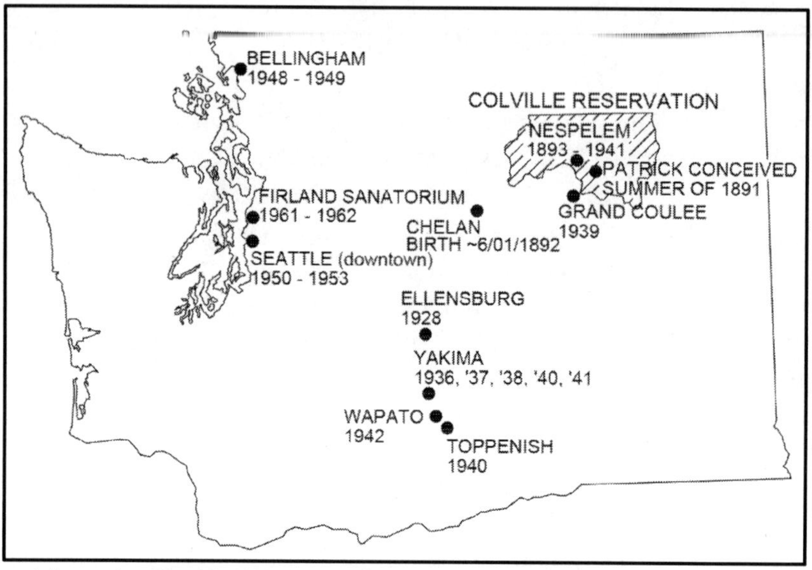

A Geography of Patrick: 1891-1962; Washington State, United States. By the author, 2025.

CHAPTER TWENTY-FIVE

Do We Have the Right Man?

At the end of Chapter 2: *The Resurrection of Patrick's Story*, I summarized the facts about Patrick uncovered by Ed Fusch with his 1985 research. In the following table I further summarize that information and compare it with what I found on Patrick through my own research. This tabular comparison may help the reader to further judge the quality of the evidence for or against a hybridization event on the Colville Reservation in the 1890s.

Of the 28 potentially provable points that Fusch mentioned, 10 (or 36%) were confirmed as consistent with the biographical history of the individual named Patrick [redacted] whom Heather Moser initially identified (points 2, 5-7, 12,13, 17, 18, 21, and 27 in the chart). On 9 points (32%) the Fusch information was not consistent with the biography of the particular Patrick that I researched (8-10, 14, 16, 19, 20 and 26).

I was not able to negate or confirm 6 (or 21%) of the remaining 10 unresolved points (1, 4, 11, 15, 22, and 28). 2 (or 07%) were inaccurate matters of remembered family history, decades old, that may be reasonably attributed to the fog of time and distance (23 and 24). One last item (04%), #3, was neither proved nor disproved (Hahissiat's abduction location from a fish camp), but was fully consistent with Colville economics and culture of the time.

	FUSCH REPORTING ON PATRICK, 1992	SOLLIE FINDINGS, 2025
1	Arms reaching about to his knees	Neither confirmation or negation
2	Abduction around the turn of the century	**Yes**, Patrick was born in the early 1890s #
3	Abduction at a fish camp near Keller, WA	Consistent with Native habits of the time
4	The abduction lasted all summer, or at least a couple months	Neither confirmation or negation
5	Resulting in the child Patrick on the Colville Reservation	**Yes**, a Patrick grew up on the Reservation at this time *
6	About 5'4" tall	**Yes**, Patrick was very short, likely about 5'-2" tall. **
7	Unusually sloping forehead	**Yes**, Patrick had an unusually sloping forehead. **
8	Large lower jaw	No **
9	Large and wide mouth	No **
10	Ears elongated upwards (peaked) and bent out at the top	No **
11	Large hands and long fingers	Neither confirmation or negation
12	Very ugly	**Yes**, not attractive. **
13	Pinhead	**Yes**, his head was small, especially viewed from front **
14	Larger than normal ears	No **
15	Extremely intelligent	Neither confirmation or negation
16	A gentle man who never beat or mistreated his wife	No, attacked wife Louise & nephew w/ knife in 1940 ****
17	Operated a ranch in the Nespelem area	**Yes** ***
18	Married easily	**Yes**, he married in 1916 at the age of 24 ***
19	Had three daughters and two sons	No, Patrick had at least 4 daughters & 4 sons
20	Died at about age 30	No, he died at a likely age of 70 *****
21	Patrick is buried on the Colville	**Yes** *****
22	Unusually intuitive card player	Neither confirmation or negation
23	Both sons died at an early age	Sons 1 & 3 died young. Son #2 died at 30 & #4 at 31 yo.
24	Mary Louise about 65 yo in 1985	In 1985, Marie was 60 yo & Mary Rose (Rosalie) was 50.
25	Madeline still living in 1985 on the Washington coast	No, Madeline Christine died in Seattle 8/14/1962 *****
26	Stella was the last born daughter	No, Stella was 2nd out of 4
27	Stella died young	**Yes**, Stella died at 25 years old *****
28	Louie, worked on Patrick's ranch ~1925 to 1930	Neither confirmation or negation

See Chapter 8: *The Birth of Patrick*

* Extensive documentation on this child was found and described in this book; See Chapter 13: *The Young Patrick*

** See Chapter 16: *Patrick's Physiology*

*** See Chapter 14: *Patrick Age 10 to 30*

**** See Chapter 23: *The Aging Patrick*

***** See Chapter 24: *The Death of Patrick*

Table 3: A comparison of Fusch's published findings on Patrick in 1992 versus author's research in 2025.

If we were to theorize that Fusch's informants were feeding him a tall tale with a beginning almost 100 years before the telling (1891 versus 1985), it would be an incredible stroke of luck for their invention to correspond with documented reality at least 36% of the time. Throw in the six unconfirmed data points along with the fish camp element and there is a potential for as much as 61% accuracy in the original tale.

I don't actually believe that Ed's informants told him the whole truth. I think they started off speaking freely with him up until the moment they noticed that this crazy white man *actually believed them*. I suspect that they then had an "*Oh,* [redacted]*!*" moment when they realized that they were revealing family secrets that should not leave the reservation. The Colville peoples had enough trouble in the white world without the Anglo invaders realizing that there was Skanicum blood on and off the Reservation. Native Americans are often very protective of their Night People relatives and they are, of course, protective of their own people. At a certain point they realized that they had gone too far in trusting this Ed Fusch. Damage control kicked in and they did what they could to conceal and distract from some of the scandal of the story by telling Ed that Patrick was a good husband who had died young.

It may be relevant to note that two of Patrick and Louise's sons died at age 30 (Ernst in 1954 and Gabriel Joseph in 1969). The Ed Fusch informant claim that Patrick himself died at age 30 (item #20 on the chart), may have been a simple confusion about people and events already over twenty years in the past.

I am absolutely convinced, through my personal research, that the journey described to Ed Fusch in 1985 of a half Skanicum, half *Homo sapiens* child born to a Colville woman almost 100 years previously was based on actual events. The kinship ties described by his anthropological informants matches very closely to that of painstakingly documented lives. I have followed the trail of Patrick's genetics through the late 1800s, the 1900s, and the early 21st century. In addition, the surviving 1928 images of Patrick show some extremely odd features (detailed in Chapter 16: *Patrick's Physiology*) that cannot simply be attributed to *H. sapiens* variability.

Patrick was a successful—actually, fruitful—genetic introgression of Sasquatch into *H. sapiens* that continues to expand through North America today. Presumably, Patrick was not the only successful relict hominin introgression into North America or into the other continents of the world, rightfully challenging our perception of what it is to be human.

CHAPTER TWENTY-SIX

Patrick's Legacy

Patrick had at least eight children through his marriage with Louise. We will have a hint in this chapter that there may have been two other children, though they were presumably childless themselves.

We left the Patrick story with his death from cardiac failure in Seattle's Firland Hospital on August 5th, 1962. I believe it no coincidence that his last known living son, Gabriel, had yet another unspecified recorded scrape with the law just three days after Patrick's death.[686]

The documents never could stick to one birthdate or age for Patrick. In three documents giving his age at death, one gives 30 (Fusch's paper), one gives 76 (the death certificate), and one 64 (the obituary below). I think that he likely had his 70th birthday about June 1st in 1962, and died four months later.

This obituary gives a last name for Marie that I had never seen before, implying a marriage #2 for her, leaving Melchor Kasco [redacted] Jr.—whom she had married in 1948—behind (though not legally divorced from yet). This second "marriage" of Marie's acknowledged in this 1962 obituary is also puzzling as this

[686] 8/081962, Gabriel [redacted]; Inmate Mug Shots and Control Cards, Washington State Corrections and Jail Records

Pacifico F., another Filipino, had married Maura De La Cruz on 6/19/1945 and appears to have remained married to Maura until his death in 1982.

The obituary for Patrick published on August 10th, 1962:

Patrick [redacted]

NESPELEM — Patrick [redacted] [redacted], 64, died Monday in Seattle after an illness of 18 months.

He was born in 1898 in Chelan and later moved to Nespelem where he married Louise [redacted] in 1918. He was a grain rancher at Nespelem before his retirement in 1947. He was a member of the Catholic Church.

Surviving are his widow; one son, Gabreal [redacted]; and three daughters, Marie [redacted], Madelyn [redacted] and Mary Rose [redacted] [redacted], all of Seattle; and nine grandchildren.

Patrick's published obituary from August 10, 1962. There are several inaccuracies in the copy, including the year of his birth and marriage and the spellings of two of his children's first names. His last name and Louise's maiden name are correct.

The last two words in the obituary should be noticed: *nine grandchildren*![687] In 1962 there were nine *living* grandchildren! These grandchildren would all have been 1/8 Skanicum and no more than about ten years older than me (born in 1959). A couple of these grandchildren may have been born after me. Some of these grandkids would have gone on to have their own 1/16 Skanicum children, giving this genetic introgression event a firm foothold in North America. In the collective sense, *we are blood relatives* of the things that go bump in the night.

The child that I warned was already in the hospital when Patrick died was Madeline Christine. She had entered King County Hospital in Seattle on August 2, 1962, three days *before*

---

[687] 8/10/1962, Patrick [redacted]; Obituary

Patrick died and she herself died on August 14th of Laennec's cirrhosis (generally resulting from long-term heavy alcohol consumption). Pneumonia contributed. The death certificate indicates that she had lived in Seattle for 13 years (perhaps leaving Bellingham and her husband, Joseph Sr., for good soon after her daughter Donna Marie had died in 1949). Her date of birth matches the early census records from the Colville. She is given the same last name as her old roommate William Oliver from the house near Gas Works Park in Seattle. The certificate says that she was married at death (presumably to the same William Oliver [redacted]), but I have found no marriage licenses at all for Madeline Christine and a relative of Willam Oliver insisted that he had died without ever having married.

Louise [redacted]

NESPELEM — Louise [redacted], 68, died Saturday in Seattle where she made her home for the past 22 years.

She was born near Chesaw and married Patrick [redacted] at St. Mary's Mission in 1917. They lived here for many years.

. Surviving are two daughters, Marie [redacted] and Mary Rose [redacted], and a son, Gabriel [redacted], all of Seattle; four brothers, a sister and eight grandchildren.

William Barnes is in charge of arrangements.

Louise's obituary from the 2/18/1963 issue of the Wenatchee Daily World. Last names have been redacted to protect any potential living relatives. Marie still carries the puzzling last name of a possible second marriage since Melchor Kasko in 1948 and the number of grandchildren has dropped from nine to eight in the ten months since Patrick's death. Louise actually married Patrick in 1916, not 1917.

Madeline Christine died at 32 years of age.[688]

Louise, Patrick's long estranged wife, truly free of Patrick at last, would herself die only six months later on February 12th, 1963, with the same fatal condition as Patrick—arteriosclerotic heart disease—for 'yrs"—a condition that may have manifested in her with chronic fatigue, chest and leg pain, dizziness, and shortness of breath. She was recorded as widowed (not divorced). Perhaps Patrick's (and presumably the family's) Catholic beliefs had prevented Patrick and Louise from divorcing despite all the trauma between the two.

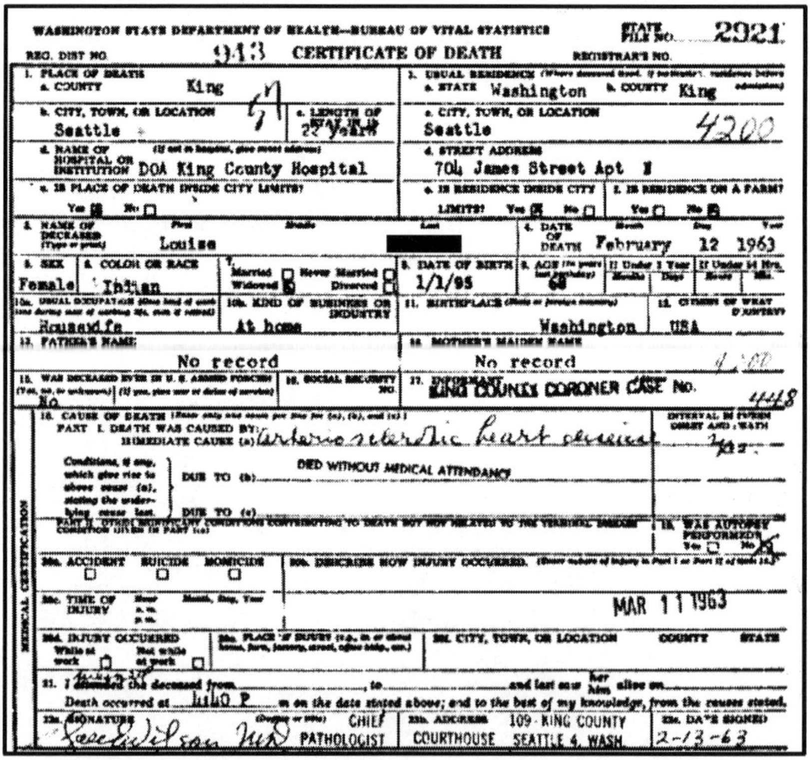
WASHINGTON STATE DEPARTMENT OF HEALTH—BUREAU OF VITAL STATISTICS
STATE FILE NO. 2921
REG. DIST NO. 913
CERTIFICATE OF DEATH
REGISTRAR'S NO.
1. PLACE OF DEATH a. COUNTY King
b. CITY, TOWN, OR LOCATION Seattle
c. LENGTH OF STAY IN THIS PLACE 22 years
d. NAME OF HOSPITAL OR INSTITUTION DOA King County Hospital
e. IS PLACE OF DEATH INSIDE CITY LIMITS? Yes [x] No [ ]
2. USUAL RESIDENCE a. STATE Washington b. COUNTY King
c. CITY, TOWN, OR LOCATION Seattle 4200
d. STREET ADDRESS 704 James Street Apt 1
e. IS RESIDENCE INSIDE CITY LIMITS? Yes [x] No [ ]
f. IS RESIDENCE ON A FARM? Yes [ ] No [x]
3. NAME OF DECEASED (Type or print) First Louise Middle Last [redacted]
4. DATE OF DEATH February 12 1963
5. SEX Female
6. COLOR OR RACE Indian
7. Married [ ] Never Married [ ] Widowed [x] Divorced [ ]
8. DATE OF BIRTH 1/1/95
9. AGE (in years last birthday) 68
10a. USUAL OCCUPATION Housewife
10b. KIND OF BUSINESS OR INDUSTRY At home
11. BIRTHPLACE Washington
12. CITIZEN OF WHAT COUNTRY? USA
13. FATHER'S NAME No record
14. MOTHER'S MAIDEN NAME No record
15. WAS DECEASED EVER IN U.S. ARMED FORCES? No
16. SOCIAL SECURITY NO.
17. INFORMANT KING COUNTY CORONER CASE NO. 448
18. CAUSE OF DEATH PART I. DEATH WAS CAUSED BY: IMMEDIATE CAUSE (a) arteriosclerotic heart disease
INTERVAL BETWEEN ONSET AND DEATH yrs.
DUE TO (b) DIED WITHOUT MEDICAL ATTENDANCE
DUE TO (c)
PART II. OTHER SIGNIFICANT CONDITIONS CONTRIBUTING TO DEATH BUT NOT RELATED TO THE TERMINAL DISEASE CONDITION GIVEN IN PART I (a)
19. WAS AUTOPSY PERFORMED? Yes [ ] No [x]
20a. ACCIDENT [ ] SUICIDE [ ] HOMICIDE [ ]
20b. DESCRIBE HOW INJURY OCCURRED.
20c. TIME OF INJURY
MAR 11 1963
20d. INJURY OCCURRED While at work [ ] Not while at work [ ]
20e. PLACE OF INJURY
20f. CITY, TOWN, OR LOCATION COUNTY STATE
21. I attended the deceased from ______ to ______ and last saw her alive on ______
Death occurred at 1140 P m on the date stated above; and to the best of my knowledge, from the causes stated.
22a. SIGNATURE [signature] M.D. CHIEF PATHOLOGIST
22b. ADDRESS 109 KING COUNTY COURTHOUSE SEATTLE 4, WASH.
22c. DATE SIGNED 2-13-63

Louise's Certificate of Death. Died February 12, 1963.

The death certificate shows Louise as a Seattle resident for 22 years (or since around 1941—about the same beginning Seattle

[688] 8/14/1962, Madeline Christine [redacted]; Certificate of Death; Bureau of Vital Statistics; Washington State Department of Health

residency also implied on Patrick's Death Certificate). She died without medical attendance as an "At home" "housewife" and was pronounced dead on arrival at Seattle's King County Hospital at 68 years old.[689]

The records show that Gabriel was transferred from the Washington State Reformatory in Monroe to Okanogan (OHC) on 4/6/1963.[690] I don't know what OHC stood for in 1963. Presumably it was the county jail where he must have been required to serve a county sentence after completing his time for state charges. Apparently this county sentence was only 60 days because Gabriel was paroled on June 3rd, 1963.

As far as I can tell, Gabriel stayed out of the reach of the law after 1963 and even pulled a marriage license at the end of 1966 when he was 28 years old, marrying Phyllis Charlene [redacted], 20, of Seattle, on January 7th, 1967. His brother-in-law, Delbert, was one of the witnesses to the marriage.[691] Presumably, Gabriel's sister and wife of Delbert, Rosalie (Mary Rose), attended as well.

Ironic and tragic are words that do not do justice to Gabriel's death less than two years later, leaving behind a 22-year-old widow. Gabriel and Phyllis were living less than two miles from downtown Seattle in the Central District. He had been working as a dishwasher and was a month short of his 31st birthday when he died.

The Death Certificate says that Gabriel Joseph died of a brain stem herniation on January 3rd, 1969 due to an intraventricular hemorrhage as a consequence of a ruptured anterior communicating artery aneurysm. Considering his personal history, for a long time I assumed that someone had smashed Gabriel over the head with a blunt object. But I consulted a person with a medical background for a second opinion and did some online research. Considering Gabriel *did* receive an autopsy, coupled with the fact that his death was not flagged as a homicide

---

689 2/12/1963, Louise [redacted]; Certificate of Death; Bureau of Vital Statistics; Washington State Department of Health

690 4/06/1963, Gabriel [redacted]; Inmate Mug Shots and Control Cards, Washington State Corrections and Jail Records

691 1/07/1967, Gabriel Joseph [redacted] & Phyllis Charlene [redacted]; Marriage Certificate, Washington State Marriage Records

or accident, I believe he simply had a sudden blowout of a major cerebral artery near the forward base of his brain (a ruptured anterior communicating artery aneurysm). This ruptured artery sent blood in and around his brain tissues (intraventricular hemorrhage), probably causing a severe headache with nausea, confusion, rapid unconsciousness, and seizures. There was so much internal bleeding and pressure on his brain that his brain stem literally pushed out of the cranial opening for the spinal cord at the base of his skull (a brain stem herniation), likely causing the additional symptoms of irregular breathing and pulse, and eventual cardiac arrest.

Simply: His brain exploded and his heart stopped.

I think death came quickly for Gabriel. No interval between the onset of symptoms and death was given, but the time of death was 10:35 PM, a prime time for dishwashers to be at work, after the dinner rush.[692]

Most likely unrelated, but the area of Gabriel's brain aneurysm is the same area in which I detected peculiar cranial structure in Patrick—the back of Patrick's brain case extended beyond the norm, low and forward, increasing his cranial capacity. Perhaps something of this odd cranial structure was inherited to some extent in Gabriel as an eventually fatal congenital defect.

Gabriel, ¼ Skanicum, the last surviving son of Patrick and Louise, was dead in 1969.

The trail of Patrick's two remaining children and his grandchildren begins to grow less distinct from here. I am okay with that. I have no desire for living relatives to be asked questions. I did find evidence that unlucky Gabriel probably left no living heirs in the form of an obituary for Phyllis's father, a Makah Indian, from 2/25/1971, only about two years after Gabriel's death. In this obituary, Phyllis is implied as living under her maiden name and still residing in Seattle. No grandchildren are claimed for her father. I have been unable to find indications that there were any children as a product of this short marriage to

---

[692] 1/03/1969, Gabriel Joseph [redacted]; Certificate of Death; Bureau of Vital Statistics; Washington State Department of Health

Gabriel.[693]

I found no other recorded traces of the Patrick line until 1973, when Ray M. [redacted] (formerly married to Stella in 1946) is buried on January 27th in Shoreline, just north of Seattle (very close to Firland). Ray M. would have been about 74 years old at his death. Interestingly, this grave location is shared with a "Patrick" of the same last name as Ray M. I speculate that this was the only child of Stella Lucy and Ray M., named after Stella's father, Patrick. This Patrick probably died young because I find no other reference to him anywhere. There are no references to Stella Lucy either between her marriage to Ray in 1946 and her death in March of 1953, almost 20 years before the death of her much older one-time husband.

As of 2015, there were no visible markers on the lonely gravesite that Ray shares with the mysterious Patrick [redacted].

Rosalie (sometimes known as Mary Rose or Mary), seventh known child of Patrick, left more of a mark on the world and successfully passed on the genes that she carried within her. I believe that she had four sons by Delbert Dee. One of those died as an infant (Daryl), but one that was born in about 1957 shows up in a picture from 1973 when he would have been about 16 years old. He looked and dressed much like I did at the time—wide polyester collar under a knit vest and way too much hair with bangs pushed to the right side. But he also has *a noticeably short neck*. This is the same presumed compressed cervical vertebrae feature that we saw in Patrick, and in his children Ernst and Marie. Interestingly, even though Gabriel Joseph did not seem to have this short neck feature, compressed cervical vertebrae apparently persisted down at least into 1/8 Skanicum blood.

Sadly, just a few months after the picture of this son was taken, his father and Rosalie's husband, Delbert Dee, died of a heart attack after a history of heart problems on August 10th, 1973. A residence address is given on the Death Certificate of Kent, Washington, but Delbert Dee died in the evening at a cafe in Omak, on the edge of the Colville. His occupation is given as sheet metal worker, perhaps associated with the growing Boeing

---

693 2/25/1971, Elwood M. [redacted]; Obituary; *Peninsula Daily News*

Company (an important company in Ed Fusch's life as well). "Mary Rose" (Rosalie) is both the surviving spouse and the "Informant" on the Death Certificate. Delbert Dee was only 43 years old at his death and Rosalie was already widowed at 38.[694]

Now Rosalie was caring for perhaps three boys so it is no surprise that three years later, in 1976, she and Joseph W. [redacted], 12 years her senior—both living in Kent, Washington—are granted a marriage license.[695] However, there appears to be a hold-up in tying the knot for these two. We won't hear more about this couple until 1981 with yet another plot twist.

As already mentioned, William Oliver [redacted] dies on September 28th, 1977. Madeline Christine in 1948 was living with William in Seattle and dies with his last name in 1962. Confusingly, William Oliver's Death Certificate insists that he was never married. His sister is listed as the Informant for that document.[696] Perhaps the sister preferred not to acknowledge her brother's marriage to Madeline Christine? Along with this confusion, I have found no indication that any children came out of this union.

We haven't finished uncovering evidence that Patrick's Skanicum father's genes gained a lasting place in North America. Next is the first evidence that one of Patrick's grandchildren (1/8 Skanicum) had married. The same gentleman (child of Rosalie/Mary Rose) that I found pictured at 16 in 1973, married at 21 years old, in 1978. His new wife is 17. Witnesses to this marriage are a young man and woman who turn out in later research to be *siblings* of the groom (other grandchildren of Patrick's).[697]

---

[694] 8/10/1973, Delbert Dee [redacted]; Certificate of Death; Bureau of Vital Statistics; Washington State Department of Social and Health Services

[695] 8/07/1976, Mary R. [redacted], & Joseph W. [redacted]; page 4, Licenses; *Spokane Chronicle*

[696] 9/28/1977, William Oliver [redacted]; Certificate of Death; Bureau of Vital Statistics; Washington State Department of Social and Health Services

[697] 1978, [redacted] & [redacted]; Marriage Certificate, Washington State Marriage Records. **Note**: as my documentation begins to have the potential to lead to living persons whose privacy *must be respected*, I am going to be deliberately vague about details, even in the footnotes.

Could this marriage turn out to be part of a revitalization of the Patrick line?

The answer to that question is yes and no. In 1980 the Mrs. had already filed for and was granted dissolution of her brief marriage to our 1/8 Skanicum short-necked male grandchild (son of Rosalie/Mary Rose and Delbert Dee). BUT the same paperwork indicates that there were *two living children* as a product of that short marriage![698] Two 1/16 Skanicum children with one likely born in 1978/'79 and the second in 1980. At the time of this writing these kids would only be about 45 years old, but possibly are each still raising their own families of 1/32 Skanicum children (great great-grandchildren to Patrick).

I promised a 1981 plot twist through Rosalie/Mary Rose. You may remember Mary Rose had her marriage license to Joseph W. [redacted] published five years earlier in 1976. But there was no subsequent wedding. Imagine my surprise on finding an "Application for Marriage License" again (apparently) granted to Mary Rose and Joseph W. late in 1981. They give the same residence address. Why the hold-up?

Well, look here in the small print: Turns out Mary Rose checked the box for "Divorced".[699] But Delbert Dee died in 1973 leaving Mary Rose, a widow with three years to go before she and Joseph W first pulled a marriage license. Wouldn't she put an x in the "Widowed" box on the 1981 form?... Unless... unless she had married soon *after* she became a widow and *before* she got together with Joseph W. Then the awkward fact surfaced that she was still legally married to this mystery man and she had to get a divorce from him before she could finally marry her live-in partner, Joseph W.

Mary Rose would still have been in her late 30s and early 40s from 1974 to early 1976 in this mysterious three-year gap. She

---

However, any footnote offered does confirm that I have found specific records to support the narration.

[698] 1980, [redacted] & [redacted]; Certificate of Absolute Divorce or Annulment; Bureau of Vital Statistics; Washington State Department of Social and Health Services

[699] 1981, Mary R. [redacted] and Joseph W. [redacted]; Application for Marriage License; Okanogan

was young enough to have a child in this time between Delbert Dee and the elder Joseph W of a presumed second marriage (out of three total).

One week after the marriage license application was again processed, Mary Rose and Joseph William were finally married in December of 1981.[700]

Writing of unfinished paperwork, it is not until March 2nd, 1982 that Marie—of downtown Seattle "cutting scrape" fame—finally successfully petitions Melchor Kasko [redacted] for divorce, almost 32 years after they had married in Seattle (and overlapping a supposed second marriage to Pacifico F [redacted] and a *third* marriage to Baptiste Casey [redacted]). In 1982, Melchor Kasko is living on Kodiak Island in Alaska. At the time of this filing, Marie is still living in Seattle at age 56, over 40 years after she had moved there from Nespelem at the age of 15.

Significantly, this divorce form indicates that there were no children born alive out of this marriage between Marie and Melchor Kasko.[701]

This new, uncomplicated partnership or marriage between Marie and Baptiste Casey did not last long, unfortunately. Baptiste died about 16 months later, at 70 years old, of congestive cardiomyopathy from hypertension and with pneumonia. He died at home in Seattle. Baptiste was a Native American from Montana, but had lived in Seattle since 1941, an iron worker before he retired. Marie is listed on the Death Certificate as his spouse and his eldest son by a previous marriage acted as the Informant.[702]

According to a Montana Obituary, Baptiste was buried in that state after being married to Marie since 1961. Of significance to this Patrick bloodline research, *no* surviving children to the marriage with Marie are listed, just two sons by a previous

---

[700] 1981, Mary Rose [redacted] and Joseph William [redacted]; Certificate of Marriage; County of Okanogan; State of Washington

[701] 3/02/1982, Marie [redacted] ([redacted]) & Melchor K. [redacted]; Certificate of Dissolution or Declaration of Invalidity of Marriage; Vital Records; State of Washington Department of Social and Health Services

[702] 1983, Baptiste C. [redacted]; Certificate of Death; Vital Records; State of Washington Department of Social and Health Services

marriage.[703]

To all accounts, Marie's long marriage (legal or in practice) to Baptiste was a happy one that turned her life around from her early days as a young teen in frequent trouble with the law in Seattle. Hers was ultimately one of the success stories of Patrick's family and much of that success may lie with Baptiste's influence.

Marie [redacted] d Aug 7 1983

NESPELEM — Marie [redacted], 59, Seattle, died Sunday at Harbor View Hospital in Seattle.

She was born Aug. 26, 1924, at Nespelem. She attended schools at St. Mary's Mission and moved to Seattle in 1939. She married Baptiste [redacted] Seattle and was a member of the Colville Confederated Tribes and the Catholic Church.

Survivors include one sister, Mary [redacted], [redacted]; and two stepsons, [redacted] and [redacted], Seattle.

Scharbach Funeral Home, Grand Coulee, is in charge of arrangements.

August 1983 *Wenatchee World* obituary for Marie, third known child of Patrick and Louise.

704

Perhaps it is no surprise that Marie did not live long after Baptiste died, even though he was twelve years her senior. Marie died at Seattle's Harborview Medical Center on August 7th, 1983—barely one year after Baptiste. She was only 58 years old (wrongly claimed as 59 in her obituary). On the Death Certificate she is noted as married to Baptiste; that she was a housewife; and that her parents were Patrick and Louise. Cause of death was gastrointestinal hemorrhage—bleeding from some type of perforation or ulcer in her digestive tract. Alcoholism can contribute, but I prefer to believe Marie had not been drinking much in her later years.[705]

Carolyn Myss writes that "biography becomes biology." In the last year of Marie's life, the story of losing her loving husband of

---

703 1983, Baptiste [redacted]; Obituary in a Montana newspaper

704 Aug 1983, Marie [redacted]. [redacted]; Obituary; *Wenatchee World*

705 8/07/1983, Marie [redacted] [redacted]; Certificate of Death; Vital Records; State of Washington Department of Social and Health Services

over twenty years manifested biologically as literally bleeding to death on the inside.

Melchor Kasko [redacted] Jr., Marie's second significant life partner, would live for more than 20 years, dying on December 2nd, 2004 and was buried in a military cemetery in the Seattle area. He was 80 years old.

Health
CERTIFICATE OF DEATH
051
146 3 08398
Mary Rose [redacted] [redacted] Female Mar.10, 1993
58 Jan.12,1935 Nespelem, Wa. No Okanogan
[redacted] [redacted] No
Widow NA [redacted] 12 0
Labor. Seamstress Factory No Amer.Indian
[redacted] [redacted] No Okanogan 39 Y. Wash. 98841
Patrick [redacted] Louise [redacted]
[redacted] [redacted]
Mar. 11, 1993 1415
[redacted] [redacted]
Gram negative sepsis 4 days
Urinary tract infection ?
Cirrhosis with end stage liver failure No No
Ella Robbins, Chief Registrar 3-12-93

Known variously as Rosalie, Rosalia, Mary, and Mary Rose—last living ¼ Skanicum child of Patrick and Louise. Died March 10, 1993 at 58 years old, over one hundred years after Patrick's birth in approximately June of 1892.

Joseph [redacted] Jr—Madeline Christine's son (Patrick's grandson at 1/8 Skanicum) by Joseph [redacted] Sr—married in 1984. Joseph Jr's mother had been dead now for over twenty years.

The last living child of Patrick and Louise's ("Rosalie" at birth), and known as Mary Rose or Mary later in life—seventh born out of eight known children—died on March 10th, 1993 on the edge of the Colville Reservation in eastern Washington. She was also 58 years old at death. There is a belief held by some in the Sasquatch enthusiast world that the Night People have extraordinarily long lives of up to three hundred years. That was not the case for Patrick or of any of his eight ¼ Skanicum children. Of course, health care and lifestyle were not good for Native Americans in the twentieth century and today they continue to struggle with health issues out of proportion to the richness of the continent taken from them.

Mary Rose died in a hospital after four days of illness with gram negative sepsis—an over-reaction of the immune system to a bacterial infection creating an inflammation response with high fever, very low blood pressure, and spreading organ failure. Broad spectrum antibiotics can reverse the course of sepsis, but even in the last five years I know of two people that died of sepsis even while hospitalized.

Mary Rose was cremated after living back in Okanogan County for 39 years. That was plenty of time to establish or re-establish strong family ties. Her father, Patrick, and mother, Louise, were identified correctly on the Death Certificate.[706]

It is interesting to note, Mary Rose lived about a year past the publication of Ed Fusch's paper with its significant section on the Skanicum hybrid—Patrick. Ed had already lived for many years very close to Mary Rose. According to the paper, Ed's informants had a good idea of where Mary Rose resided. I suspect Ed did engineer a meeting with Mary Rose, but only under a vow of secrecy, which to his credit, he honored beyond his death.

In Ed's 1992 paper, the erroneously identified "Mary *Louise*"

---

[706] 3/10/1993, Mary Rose [redacted]; Certificate of Death; Washington State Department of Health

is described as having a physical appearance that is "relatively normal."[707] Possibly Ed wrote this from personal observation. It is hard to imagine that this enormously curious man—who had researched in Hawaii, Africa, and Kentucky—would not make a point of finding Patrick's last known child, knowing that she lived *less than ten miles* from his schoolhouse home. It is even possible that the flurry of visits by the French professor, Dr. Jean-Paul Debenat, in 1990, 1991, and 1992 had something to do with Mary Rose and that she was a pivotal inspiration for the publication of his 2009 book, *Sasquatch/Bigfoot and the Mystery of the Wild Man: Cryptozoology and Mythology in the Pacific Northwest.*

We are almost done exploring the legacy of Patrick's descendants before we must stop the digging or risk getting too close to living, breathing people, carrying in their cells what many may think a biological impossibility.

Mary Rose [redacted] Pub- 12-March- 1993

[redacted] — Mary Rose [redacted] 58, of [redacted], died Wednesday, March 10, 1993, in [redacted] at [redacted].

She was born Jan. 12, 1935, to Patrick and Louise [redacted] at Nespelem. She was born and raised in Nespelem, then moved to the Seattle area. She returned in 1978 to [redacted] where she had lived since.

She graduated from Chemawa Indian School in Salem, Ore. She married Delbert [redacted] in 1954 at Seattle. She was a member of the [redacted] Catholic Church and the Colville Confederated Tribes.

She is survived by four sons, [redacted]; four daughters, [redacted]; and 16 grandchildren. She was preceded in death by her parents, her husband, her son, Daryel; and nine brothers and sisters.

Arrangements are by [redacted] Chapel of [redacted].

Wenatchee World 3/12/1993 obituary for Mary Rose (Rosalie), the seventh known and last living child of Patrick and Louise.

Mary Rose's obituary above carries bombshells: Eight living children with *six* different family names. And *16 grandchildren*![708] The number of grandchildren is the biggest

[707] Fusch 1992a, page 20

[708] 3/12/1933, Mary Rose [redacted]; Obituary; *Wenatchee World*

bombshell. These would be great-grandchildren of Patrick. If still living, they are all old enough now to have had children of their own for a few years already, and perhaps to have more children for a few more years into the future!

Whomever wrote the obituary also thought that there were a total of ten children in the Patrick-Louise family. I have not found two of these children. Did they die young, leaving no records? Were they more recent half-siblings from Patrick's itinerant years (starting about 1940), when he was in his late 40s? If these other two ¼ Skanicum kids ever existed (and possibly still living), did they have children of their own?

Most likely, whomever was doing the tabulating just got thrown, as did I, by the name changes of some of the siblings over the course of their lifetimes. I strongly suspect that this figure is owing to a double-count of a couple of the daughters.

Considering Patrick's genetic legacy and ignoring these two mystery siblings: None of his four known boys had recorded children. Of the four girls, three—Stella Lucy, Madeline Christine, and Rosalie (Mary Rose/Mary)—successfully bore children. I suspect and assumed in my projections that Stella Lucy's only child died young, without a chance to have kids of his own.

You will see in Table 4, *Known and Predicted Propagation of Skanicum Genes*, I attempt to project the spread of Patrick's Skanicum genes all the way through seven generations removed (1/128th Skanicum) from Hahissiat's abductor, to about the year 2240. There was much guesswork in these calculations, but I tried to err towards the low side. I used current and past fertility figures (the average number of live birth children a woman has in her lifetime in the United States), where available. For projecting future decades of the growth of Patrick's lineage, I used 1.7 as an approximation of the current fertility number (replacement level is 2.1 to account for infertility and mortality).

Rosalie (Mary Rose/Mary) was gold for this genetic introgression event, giving Patrick and Louise as many as nine grandchildren and 16 great-grandchildren (at 1/16th Skanicum each) as of her death in 1993. There certainly may have been more great-grandchildren of Patrick's in Rosalie's line, after her early death, or even in the family tree of Madeline Christine.

Marie appears to have been childless despite three marriages.

Stella Lucy gives the hint of only one child (also named Patrick) that likely died young and was buried in the family plot of Stella Lucy's brief (?) husband, Ray M [redacted] of Seattle.

Madeline Christine had at least four children. Unfortunately, the two girls died young. At least one of her two boys born in the mid 1940s, seems to have made it into adulthood and marriage and may have produced great-grandchildren for Patrick. Their prime procreation years would have been the mid-1960s to about 1995. Any offspring born during those years would have enjoyed relatively high standards of medical care with a likelihood of surviving to have their own children.

With all that in mind, I used 4 only as a conservative estimate of additional concealed grandchildren from the three child-bearing daughters.

In total, I have unearthed 39 documented individuals with Skanicum blood from this one 1892 abduction event from Patrick at half Skanicum, through 16 of Patrick's great-grandchildren as of 1993 (at 1/16 Skanicum). Of Patrick's grandchildren at 1/8 Skanicum, I estimate that at least four are still living and not much older than I am. The family stories they may be able to tell would be priceless.

It is worth again noting that the ability of Patrick to have children of his own implies that this hybridization event was between two closely-related "species" with the same number of chromosomes. At least, this conclusion follows from our current understanding of genetics and how nuclear DNA functions in reproduction. Patrick was certainly not infertile such as is the case with a mule coming from the pairing of a horse (with 64 chromosomes) and a donkey (with 62 chromosomes).

I consider my work in Table 4 to be a low estimate of the possible numbers of Patrick descendants now and into the future. Using fertility and mortality rates, I project that there are at least 46 individuals alive *today* carrying Colville Skanicum genes. 27 have already died. Without trying to trace living relatives, I have found indications of the Patrick line in four states, including east of the Rockies. The first of the seventh generation removed from

Hahissiat's abductor are likely now coming into this world. In Native America, the seventh generation is often a significant marker and represents a long-term perspective for the benefit of people and culture. The promise of the seventh generation is that of the end of dark times and of new beginnings.

Note: Allow those of Patrick's lineage to live in quiet and privacy. It seems likely that most people would not appreciate being told that their ancestor was a Sasquatch. "Proof" to the Sasquatch community is a potential death for those whose lives may already be a daily challenge. And please, let the dead rest in peace.

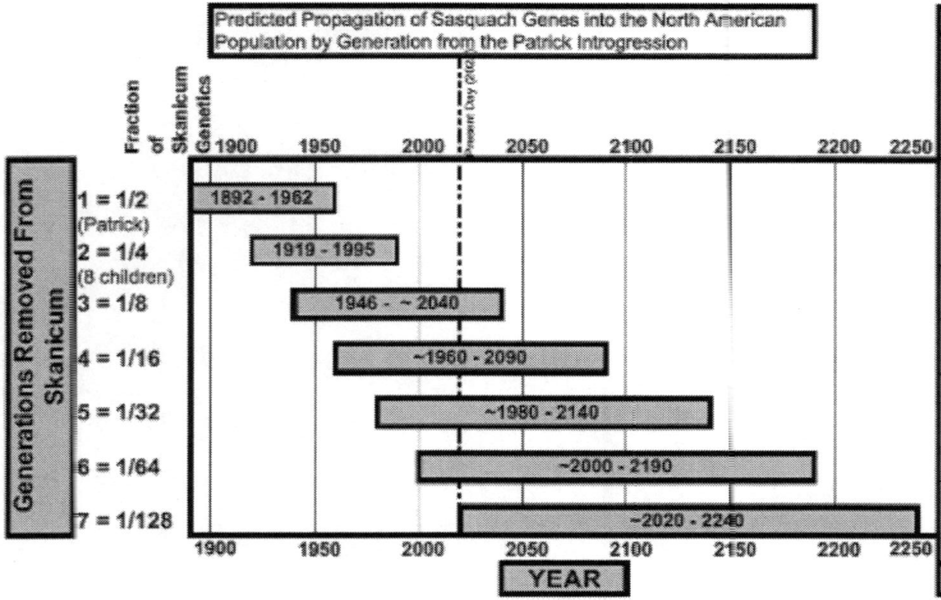

TOTAL INDIVIDUALS						
	ESTIMATED*					
KNOWN	INFERRD ADDITIONAL	ALIVE TODAY	DECEASED	TO BE BORN	W/ NOVEL GENETICS	GENERATIONS REMOVED
1		0	1	0	1	1 = 1/2 (Patrick)
8		0	8	0	8	2 = 1/4 (8 children)
14		** 4	10	0	14	3 = 1/8
16	*** 4	**** 16	3	1	20	4 = 1/16
	30	20	4	6	30	5 = 1/32
	43	5	1	37	43	6 = 1/64
	58	1	0	57	58	7 = 1/128
39	135	46	27	101	174	
TOTALS						

*Presumes fertility rate 1960s=3.0; 1970s=2.0; 1980s=1.8; 1990s=1.9; 2000s=2.0; 2010s=1.8; 2020s on =1.7.
Offspring beginning at 20 yo; ending at 50 yo; death of oldest at 80 yo.
**Estimate based on half of those known to be alive in 1993
***Estimated through Madeline Christine's two known surviving sons
****Based on ~80% of people born in the 1960s in the United States surviving to the present

N. Sollie 2025

Table 4: Known & Predicted Propagation of Skanicum Genes

TWENTY-SEVEN

After Patrick

Investigating the Patrick story, I have had to face my own biases and assumptions about Sasquatch. As a youngster in the early 1970s, I was electrified by the Patterson-Gimlin film. As a teen I fell into the cliche pattern of reading all that I could find, including the excellent John Green books. Growing into a young man in the '80s, I was squarely in the elusive-mountain-ape camp. Even then, there were hints around the edges that the mountain ape theory did not fit all accounts, but in those days the exceptions were outliers and still possible for me to ignore. I collected books and newspaper articles, and even took a solo camping trip about 1980 into the Washington Cascades, specifically to look for Sasquatch. I had inexplicable experiences on that first trip: A muddy footprint in front of my tent one morning and strange branch-breaking sounds that seemed to be encouraging me to leave. Another experience from that trip that I ignored for years: A sudden gentle but insistent need to leave the area as quickly and as directly as possible. And leave I did.

In 1984 I read a newspaper article about someone finding huge footprints under a bedroom window on the outskirts of *Bellevue, Washington* of all places. Bellevue is now a little big city, but even in those days it was a bustling and growing center of commerce and humanity. For an elusive mountain ape to hang out in a subdivision, looking through bedroom windows, made *no sense*! And unbelievable, yet seemingly sincere reports were cropping up in such peculiar places as Iowa, Indiana, and even New England.

I knew something must be wrong with the assumptions behind the twentieth century Bigfoot research, but I couldn't make the leap of understanding that all of the old guard—Peter Byrne, Rene Dahinden, Green, Grover Krantz, John Napier, Sanderson, etcetera—might have been wrong in their informed belief that the remote mountains were populated merely by a clever North American gorilla.

After reading that Bellevue report I stepped away from actively following and "researching" the phenomenon, right through my own possible (and unrecognized) sighting over 20 years later in Illinois. In the late Fall of 2011 I moved back to Alaska and found myself with time to spare as I waited for seasonal employment to begin. But it wasn't having time to spare and the internet that brought me back to Bigfoot.

It was a dream.

I know, I know! Reading about someone else's dream is about as boring as it gets, but this dream was a big part of the writing of this book, so I am going to bore you with it.

In the dream, I had been guided by another man to the rim of a canyon. We crouched low, trying to stay concealed, as he directed me to look down towards the far side of the bottom of this narrow, rocky drainage. Below us was the horizontal slit of a cave in a cliff face. In front of this cave, or rock shelter, on the long rocky ledge that served as the entry, stood ten pairs of Sasquatch all in a row, male and female, male and female...

My guide quietly explained to me that these were pairs of Sasquatch mated for life. He indicated to me that the Sasquatch were all living here together for the winter, but each pair (with any children they might have) would scatter into the surrounding country when warmer weather came.

In my dream I was astonished at this information. I whispered back, "If what you say is true, then everything that we thought we knew about Sasquatch is wrong!"

With that dream, I finally woke up to the understanding that I needed to allow new ideas to get a toehold if I wanted to gain any real understanding of Sasquatch. I decided to get back to learning more, and oh, boy, there was so much out there now in the wide, wild world of the internet and in the books being published. I began to consider what I was hearing with a more open mind, beginning with the core lesson of my dream: Sasquatch are intelligent beings that remain elusive because that's the way they like it.

This new perspective has allowed me to dig deeper and to reconsider some of the odd material in witness and personal experiences and also in the "woo"—as non-normal events and manifestations are often called. My change of heart and mind drew me to attend that conference in 2019 in Kennewick, Washington and to buy that two hundred dollar book from the Russian, Igor Burtsev. Inside I found the Patrick story. I noted Dmitri Bayanov's wish that someone in North America should take the Patrick story seriously enough to look into the evidence. It took Heather Moser only a few hours of research to convince herself, and me, that there was a real story here: One that points to an actual abduction that bore an actual male child who grew up, got married, and had children. And some of those children had children. And so on and so on...

This book may be the first widely available thoroughly detailed account of the Patrick story, but the knowledge of this saga is already out there. In the chapter on Ed Fusch, I relate how Rhettman A. Mullis Jr. of Bigfootology[709] was able to interview Ed on video in 2012. They talked about Patrick and his children. Paulides wrote about the Patrick case and the potential for obtaining DNA samples from living descendants.[710] The 2015 Burtsev book is another example of the seeds of the information already published.[711]

I am certain that there are others out there who have followed Patrick's trail through the years and are convinced that we have, in Patrick, a genuine genetic introgression event in North America. Others no doubt hesitated, as did I, to publish this information specifically out of the terrible weight of responsibility

---

[709] https://bigfootology.yolasite.com/

[710] Paulides 2017, page 78

[711] Burtsev 2015 pages 305-307

that living descendants might come under scrutiny and harassment of others eager to prove... something. Prove what? Perhaps to prove to others that *they* haven't been chasing unicorns for decades, that *they* were right all along and that the teasers and mockers were wrong? They might think if they have puzzling nuclear DNA coding in hand, or anomalous bones gathered from some diabolical midnight excavation, that the world will finally notice them and give them the respect they have been missing.

Unfortunately, the world does not work that way. Provide inexplicable DNA and the world will just say it cannot be, thus it is not. Provide some bones and they will be described as a misidentification, a hoax, or a university testing lab will simply lose the evidence.

Belief trumps logic. Money trumps logic. Power trumps logic. Change comes to entrenched anthropology—as they say—one funeral at a time. The dug-in old guard pass on and new minds and new ideas rise up to take their place.

Or sometimes a dream or some other startling event knocks people out of their mental ruts and they begin to look at things in a new way.

This book is not *proof* that the hybrid Patrick ever existed nor proof that he has transmitted Sasquatch genes into the general population through seven generations now in 2025. What this book can claim is to be a chronicle of hundreds of hours of careful research that provides convincing, anonymized corroboration of the story that Ed Fusch was first told on the Colville Reservation in 1985. Some things in that initial telling to Ed were wrong, but many elements match up exactly in records now available to anyone. There are enough factual matches, that for me—someone who took the time to personally uncover the facts—accepting that the hybrid lived and breathed and successfully passed on the Sasquatch genes that he carried from his mother's abduction is a rational act of logic, not belief.

I am convinced. I do not expect you to be convinced. You must come to your own knowledge, in your own way.

My own conviction that Patrick was real leads me to yet another enormous mystery.

I know of an event that took place in Denali National Park in Alaska sometime in the last twenty years near the Teklanika Rest Stop at mile 30 of the only major park road. This road winds west from the Park Entrance area, towards the historic Kantishna gold mining district. Two people were walking on the gravel road west of the rest stop in the quiet of late evening, but it was still the light-filled Alaskan summer—the visibility was good. They noticed something large moving on the broad glacial-river gravel bar below.

The Teklanika River gravel bar in Denali National Park & Preserve—scene of a Sasquatch moving quadrupedally. The thick Teklanika Forest approaches the far side of the river. © 2011 Alexander Sollie

They tried to identify the hair-covered creature lumbering on all four legs. The figure gave the impression of great bulk and weight, but it was hard to tell just how big it was at a distance. There was something odd about its shoulders being so much higher than its hips. Many times they had each seen the large mammals of the Park: Moose, wolf, caribou, grizzly, Dall sheep, lynx, black bear, coyote. They quickly ran through the short list and realized that none of those animals matched what they were seeing.

One said to the other, "That's a Bigfoot."

"I think you're right."

Confused, they watched this massive primordial creature work its way deliberately across the glacial gravel within sight of the road and the bridge across the river a mile upstream. It gave no sign of noticing them and the two preferred it that way. They watched it for a couple of minutes before it disappeared into the brush across the river. They had not realized before that night that Sasquatch sometimes moves on four legs. They decided it would be best to head back to the rest stop.

This is the mystery to me: How does the animal those two witnesses saw that night successfully mate with a Colville Indian and produce a hybrid child who grows to have children, secure a Land Patent signed by President Woodrow Wilson, attain United States citizenship, register for two drafts, and sign up for Social Security? The hybrid Patrick was undeniably human.

Doesn't that mean that the massive primeval creature moving on four legs through the harsh Alaskan wilderness was also human?

How can that be?

That is a question I still wrestle with and I don't blame you if that is precisely the sticking point that you face now. I do not have a *good* answer for the question. But I do have *an* answer that inevitably challenges and expands the idea of what it means to be human. In researching the anthropological record of the *Homo* genus line, I came across some hints and leads that will be followed-up in Volume II, which may be summarized as: An evolutionarily plausible explanation for the existence of Sasquatch.

However, there is still the *woo* to account for. It's not enough to explain Skanicum size, quadrupedal ability, diet, morphology, and lifestyle. How does one explain the special abilities that are described by so many earnest witnesses? In Volume III I will explore that topic and hope to give us some room to consider the extraordinary as accessible, in remarkable circumstances, to many of us.

*

In this book of examining the life of the hybrid Patrick, we have exposed truths and surprising facts.

One: There is a centuries-old belief in North America that hairy giants living in wild places occasionally take women *and* men, for food and for mates.

Two: Perhaps most significant of these facts—whatever abducted Hahissiat Kolockun in the summer of 1891 had to have been in the human family and closely related to *H. sapiens* to successfully impregnate Patrick's mother.

Three: The "type specimen" that biologists normally require before formally recognizing a new species probably was examined late in 1968 when Belgian scientist Bernard Heuvelmans and biologist and author Ivan T. Sanderson spent two days on a Minnesota farm studying a hair-covered hominin frozen into a block of ice. Heuvelmans went so far as to publish a paper on this find, suggesting the taxonomy, *Homo pongoides*: Human that resemble great apes.[712] A reasonable description.

Four: The extraordinarily lucky find of two photographs of Patrick when he was about 35 years old reveal some genuine oddities about his head and upper body. These traits may have been inherited from his Skanicum father, but hybrids often have morphological features not shared by either parent. These oddities in Patrick include: A radically short neck (implying compressed cervical vertebra); a steeply sloping forehead; a small head but with "normal" cranial capacity maintained by a skull expanded backwards and the brain volume pushed lower; ears oddly low and rotated sharply backwards at the top; the lack of a chin (no chin is a trait of early *Homo* species); and narrow shoulders. A few of these traits are visible in two out of the three children of Patrick who left behind photographs. One of Patrick's grandchildren also appeared to have an unusually short neck.

Five: Many aspects of reported Sasquatch food sources, culture, shelters, range, and technologies are similar to those of *H. sapiens* hunter-gatherer societies.

Six: Hybridizations can result in offspring either substantially bigger or smaller than either parent. While males may carry genes

---

[712] Heuvelmans 1969

intended to maximize the size of their offspring, the females of any species may carry growth-limiting genes to cancel or dampen the effects of these growth-maximizing genes in their male mates. This growth-limitation may work in two ways; 1) to shield the mother from the damage of a huge fetus, and 2) to protect her other normal-sized offspring from the potential of physically larger competition embodied in new over-sized half-siblings. A Skanicum mating with a "tiny" human woman likely is not going to result in a medium-sized Skanicum because female *Homo sapiens* carry growth-limiting genes. Hybrids are not the *average* of the two parents. The physical form of a hybrid may not appear as expected.

Seven: Perhaps Sasquatch is itself a product of hybridization in which there was a lack of growth-limiting genes in one species, coupled with growth-maximizing genes in the second species. If Sasquatch is itself a hybridization of two or more species, it may be *bigger* than any of the parent species.

Eight: Hybrids are very common among primates and are established fact even in the *Homo* line. *Homo sapiens* today carry traces of introgression events with Denisovans and Neanderthals and there are now indications of yet a third archaic hominin lurking in our genes, as yet unidentified.

Nine: Patrick's sperm was quite viable, with eight known children born over a span of about 19 years. I estimate that there are 40-some descendants of Patrick alive today as the seventh generation removed from the Skanicum abductor begin to be born.

Ten: Despite every reason to die young, Patrick lived for about 70 years in very harsh conditions. In contrast, his eight children had an average age at death of only 34 years old. His two oldest-lived children both died at 58 years old. Patrick may have been an example of "hybrid vigor", but his children apparently were not so endowed

CHAPTER TWENTY-EIGHT

60,000 Years of War: Healing the Divide

"Could it be that the accepted view of Indigenous Australians simply wandering from plant to plant, kangaroo to kangaroo, in hapless opportunism, was incorrect?"

Bruce Pascoe 2018, page 2;
Dark Emu: Aboriginal Australia and the birth of agriculture

The elusive mountain ape, Bigfoot—wandering from plant to plant, deer to deer; living by instinct—somehow outwitting us all.

As the years since Patty pile up, this belief becomes harder and harder to maintain. Pascoe goes on to write that, "by adjusting our perspective by only a few degrees, we see a vastly different world…"[713]

I suspect the story in Australia was the same story that played out earlier in North America: A great wave of death through introduced disease swept ahead of the invading Europeans, like ripples after a rock is dropped into a pond (the Plymouth Rock of

[713] Pascoe 2018, page 3

1620?).[714] These ripples of death—and the ensuing societal and cultural destruction to the existing residents—raced ahead of European exploration and conquest as epidemic after epidemic swept across the indigenous populations of the continent. Behind the catastrophic wave of disease, a veritable *terra nullius* (nobody's land) of silent landscapes; though the signs of former use and habitation were scattered everywhere. Among the indigenous devastated by this disaster: Skanicum, Bigfoot, the Hairy Man, Boss of the Woods, the Night People, Nahganne, Wooly Boogers, Sealatiks, Skunk Apes, Nantiinaq, the Bugwayjinini, Sasquatch, etc. They were and are known by many names. Perhaps hundreds of separate tribes and even a handful of separate species—our elder brothers—staggered with the carnage.

What played-out Down Under was a later repeat of Manifest Destiny in North America:

> Colonial Australia sought to forget the advanced nature of the Aboriginal society and economy, and this amnesia was entrenched when settlers arrived after the depopulation of whole districts found no structure more substantial than a windbreak, and no population that was not humiliated, debased, and diseased… It is no wonder that after 1860 most people saw no evidence of any prior complex civilisation.[715]

The friend of our Ed, Dr. Jean-Paul Debenat, pointed out that, "It would seem that sapiens is afflicted with an aggressivity that exceeds its purpose."[716] And Higham gives us an comparison from ancient history for this recent invasion:

> It seems that only after 50-60,000 years ago do we find humans spreading widely across all parts of Eurasia, Southeast Asia and greater Australia. One reason why humans failed to

---

[714] Koch et al 2019, page 30, estimate that by 1600 (*before* the Pilgrim landing at Plymouth Rock in 1620), 55 *million* indigenous modern humans had died in the Americas as a result of introduced disease.

[715] Pascoe 2018, page 11

[716] Debenat 2009, page 249

> settle more permanently [earlier] might be competition. Perhaps the presence of Neanderthals and other human populations in number was a key reason behind this… We see [modern] humans from this time onwards behave rather like an invasive species…[717]

This "competition" may as well be called a war. I believe that *Homo sapiens* have essentially been at war with the many "relict hominids" throughout our continents for 60,000 years. Just as the Anglo newcomers cannot truly assimilate onto this Turtle Island before we make peace with the Native Americans, the Alaskan Natives, the First Nations, and the Hawaiians, we also need to make peace with the Night People. This peacemaking is going to be a long, bumpy road because there is so much hurt and anger between our tribes after 60,000 years of conflict. There are small groups of Sasquatch, such as I have brushed against in Colorado, who are trying to make contact and establish an understanding and peace between our peoples. Then there are the kinds of groups I encountered in western Massachusetts who *perhaps* politely suggested to me through a psychic intermediary that I should go away and look for Bigfoot in the "land of my ancestors." According to Robert Kryder, there is or was a group of young male Sasquatch in New Mexico who build tall log barriers across trails overnight to keep the Hairless Ones out. And there are places in the Northwest Territories, Yukon, and the state of Alaska where trespassing is potentially or actually life-threatening.

The Western World has lost its recall of these 60,000 years of war through its cultural amnesia. I do not think the Night People have forgotten. And they are returning to lands once lost. Their numbers have been growing, it seems, since the 1950s. Sasquatch are reinhabiting their old territories. Steve Isdahl reports that within his hunting and trapping lifestyle, it is more common to hear of *Sasquatch* encounters than of cougar encounters.[718] People who have lived quietly for years in their cabins in the woods suddenly find themselves sharing their land—but now living

[717] Higham 2021, page 26
[718] Isdahl 2021, page 19

locked tight in their homes at night and no longer venturing into the "Back Forty." They seek, clumsily, to adapt to the presence of these enormously strong creatures that always seem to be a step ahead of efforts to control them in any way.

As Dr. Meldrum pointed out, skirmishes with this *other* have left names on our landscapes: Ape Caves, Spirit Lake, Skookum Meadows, Booger Hill, Booger Hollow, Devil Canyon, Diablo Mountain, Ape Canyon, the Headless Valley...[719]

Is it possible for us puny ones to coexist with the hidden goliaths inhabiting the nights and the woods?

It is easy for us to focus on our inability to trust *them*, but this distrust goes both ways. As Lapseritis points out, "...most of them still do not want to bother with the kindest of us... they find us difficult to trust."[720] He goes on to suggest that common ground between our peoples may be difficult to find: "There is absolutely nothing we have that they want except respect. We have more to learn and gain from them as living beings than they do from us!"[721]

From a man who has spent more than fifty years thinking deeply about this subject of race relations since his string of noisy encounters in the Sierra Mountains in the early 1970s, Ron Morehead writes:

> Bigfoot has consciously decided to inhabit the wild, rarely interact with so-called "civilized" humans, and they have no reason to become like us. These beings do, however, seem to find us interesting and observe what we do. To prevent exposure, utilizing their flexible vocal mechanism, they have become experts at diversion and mimicry. They have restricted our exploitation and adapted a life of stealth; occasionally being seen, but more often, just peering through the dense forest and observing our behavior—probably wondering why we are so inept with nature.[722]

---

[719] Meldrum 2013, page 113
[720] Lapseritis 1998, page 143
[721] Ibid., pages 195-196
[722] Burtsev 2015, page 337 quoting Ron Morehead

This curiosity with us—and the willingness for Sasquatch populations in some areas to brush the edges of human society—suggest a potential for understanding or communication between our peoples. Higham, in describing what he believes was a past period of intermingling between different *Homo* lines, unintentionally suggests a new way of thinking in the present day: "Imagine the opportunities inherent in meeting a completely new group of people who were doing things a little bit differently, more creatively, perhaps better than your own group…"[723]

Interaction is not impossible; it has already been achieved. Short writes of how the Klamath Indians of California were able to communicate with those that they called The Quiet People, welcoming them to their Jump Dance, the giants coming out of the woods with baskets of acorns for trade.[724] Short again, recounts how, before the Europeans, the Assiniboine would make their winter camp at the base of the Rocky Mountains. Per Short's source:

> ...for several winters the Big Man of the mountains came to her people's camp. They had a teepee for him. He mingled with the tribe and apparently got along very well with them. They could understand each other well and the big man told them many things. In the spring when they would leave to continue their travels, the big man would return to the mountains. This went on for many years, and occasionally he would bring others of his kind with him. When settlement of the area [by whites] became the norm, his visits became less and less. One season he never returned.[725]

There is change in the air and a need to try new approaches between the Big and the Little peoples. "Owl Man," a frequent contributor to Steve Isdahl's *How to Hunt* podcast, tells of a prophesy among The Big that the time of a "Great Knowing" is fast approaching. Owl Man doesn't know what this Great Knowing is, but says the prophesy is a source of much fear among

[723] Higham 2021, page 236
[724] Short 2024, page 90
[725] Ibid., pages 118

the Sasquatch. Noël tells of one of his informants, Denise from Maine, who is able to communicate telepathically with a female Sasquatch, receiving this message eerily similar to reported messages from extraterrestrials:

> More will see us. More will hear us. It is time. People with good heart will hear us. It is time. The earth is hurting. We need to communicate.[726]

Let's say you, the reader, were willing to risk your own life to reach across this great divide between peoples and begin the process of healing. How would you approach the challenge? Alley suggests that "Rather than loading rifles or engaging in persistent hunting to run down tracks, it may be just as profitable for individual researchers to proceed slowly, learning from the observation, recording signs and behavior represented..."[727] Noël suggests success might come from not trying so hard: "Don't be too focused on finding Bigfoot. It may appear to look to them like you are hunting them."[728]

Wiitala has some basic rules for how he conducts himself in the woods. He does not howl or whoop or make tree knocks—though he may respond to a vocalization or knock he believes was directed towards him. He does not chase Sasquatch or try to find them in the woods.[729]

Grossinger described one account, though unintentionally humorous, of how opportunities for friendship may present themselves by chance to those who can seize the moment. He tells of how one of his witnesses, while driving across the wild distances between the northern towns of the Yukon in the early evening one October night, saw someone standing near the road. As is the custom in that country, particularly among First Nations people, the driver slowed and pulled alongside the dark figure to offer him a ride. Focused on his driving, he did not realize until he looked out of his passenger window that the pedestrian was a

[726] Noël 2019, page 39
[727] Alley 2003 (2018 edition), page 303
[728] Noël 2019, page 233
[729] Wiitala 2021 page 55

short Nahganne! Startled, the driver pulled away quickly.[730] But what would have happened if he had rolled down his window and called out a greeting? Realistically, I think the Sasquatch would have been in the nearby woods before the window glass began to move, but I also think this likely adolescent might have left the encounter thinking that maybe some of the noisy Little People weren't all that bad and be more inclined to take a chance with them in the future.

Lapseritis has an even more substantial story of kindness bridging the gap, although it rests on the controversial "mind-speak" that is reportedly associated with some Sasquatch. A Native American in Washington state told that author about one night around 1950 when he and two of his siblings had witnessed his grandmother passing out her smoked salmon to a group of 11 Sasquatch ranging in height up to 12' tall! The witness said that two of the shorter Sasquatch (children) went into the shed with his grandmother and helped her distribute handfuls of salmon to the others. Later, the grandmother told him that, "...the Sasquatch were her friends that spoke to her 'in the mind' and said the Sasquatch were starving during a difficult winter and had asked her for food."[731]

Many find that music helps to bridge the gap, or, rarely, loud music annoys them. I like to play the harmonica in the woods, although I haven't noticed that I have attracted attention, good or bad. Although that was in the area where I was supposedly advised to go look for Sasquatch in the land of my ancestors. Wiitala describes how when he finished playing the guitar one night in the woods, he heard a disappointed "Aw!" from a suspected juvenile female. He describes another event when he was playing Pink Floyd on the guitar. Between songs Wiitala could hear an adult male talking to his sons. In one case, he could hear the females in the woods on one side of him and the males on the other. Wiitala had modern human friends with him and they found themselves increasingly comfortable with this extraordinary experience. They felt that the listeners in the dark

[730] Grossinger 2022, page 43
[731] Lapseritis 2011, pages 23 & 24

woods were somehow radiating a "calming" effect.[732]

The most mind-bending account that I have heard so far—without being able to write it off as fantasy—is what I call "The Sponge Bob Square Pants Problem." This comes to me through episode 419 of Wes Guermer's *Sasquatch Chronicles* under the innocuous title of "I would not believe it if I didn't see it". This episode must be sampled just to begin to let the implications truly sink in (the link to this episode is in the footnotes of page 361). I will summarize the account as relevant to this discussion of healing the divide between our people and their people.

It has been years since I listened to the episode, so pardon me, Wes, if I get some things wrong.

A young man, in the southeast United States somewhere, has a Sasquatch encounter in the woods. It starts when he sees a young Sasquatch hanging by his ankle wedged in the crotch of a tree. The cute hairy kid is crying for help. After an internal struggle, he is about to try helping the pathetic creature when a very agitated adult female Sasquatch (screaming) and a teenage female Sasquatch arrive. After a few moments of shocked staring, our witness leaves quickly. Sometime later, he returns to the same area, trying to get some understanding of what he has seen, and an older "normal" lady approaches him with mysterious specific knowledge of what he had experienced that previous day. She invites him to stop back at her rural house. He stops at the house later that day as it is growing dark and he is invited in to chat. The house smells like animals and has a few holes punched in the drywall. She explains that "they" have bad tempers sometimes. He realizes that "they" are upstairs. Muffins are baking in the oven, helping to cover the animal smell. "They" come downstairs. They are two young Sasquatch (the witness was used to thinking of them as "Haints"). They approach warily, but their normal fear is overcome by the promise of muffins fresh out of the oven. The witness is battling his own fear, but the kids look kind of cute. Then the naked teenage female Sasquatch comes into the house as the muffins are being given out. Contrasted with the older screaming female whom he saw in the woods, this one is very

[732] Wiitala 2021, pages 27 & 28

human-looking, just with a lot of hair. He implies that her breasts are very nice looking and that he must have stared too long because, without turning towards him, the naked teenager begins growling. The witness stops staring. It grows dark outside. He asks some questions and the lady tells him things like how she considers them family and that the kids are her kids too. She says "they" trust him because he was going to help the youngster that day. Sometimes all the grownups will be away for three days at a time hunting up to fifty miles away, leaving her to watch the kids. The lady explains that she buys horse feed as food for the extended Sasquatch group and that they stack the heavy bags for her in the barn. She says then, come see *this*. She takes him to a window looking out into the dark backyard. There is a television set pointed into the woods. He can see shining eyes staring at the lit television screen. The lady tells him that their favorite show is "Sponge Bob Square Pants." This witness can't believe all that he is seeing and hearing. Then there is a minor commotion. Maybe the teenage female comes back inside and says something to the old lady. I can't remember. But the old lady says, Oh! Big Daddy is here. I have to go talk to him. The old lady goes outside. He can see her through an open window. She is a tiny figure next to an *enormous* muscled male. The witness has never seen anything so big in his life. Then he sees and hears that the old lady and this frightening hulk are talking somehow. He can hear them. The old lady is speaking some other language with pops and clicks in it. The witness by this point can't really take any more of this mind-blowing stuff. When the old lady comes back in, he politely excuses himself and leaves, never to return. He did not want to run into that giant male *ever again*.[733]

You have to hear the narration for yourself. I believe the story for a number of reasons, much like I believe the Patrick story. Listen for yourself.

Revelatory to me—this instance of two vastly different peoples truly getting along and working together. If an old lady can do it, I think the rest of us can too. Perhaps key was her willingness to

---

[733] Sasquatch Chronicles 2018 episode 419; https://sasquatchchronicles.com/sc-ep419-i-would-not-believe-it-if-i-didn%e2%80%99t-see-it/

accept and help them on their own terms and to love them as family. This experience can be repeated, in other forms, though it is to be expected that the relationship will be uncomfortable and life-changing, as it was for the old lady.

Alley concludes, "It now seems all the more likely that reliable data on wild behavior in sasquatches will by necessity come not from lettered professionals examining sasquatch remains, but more likely from coincidental observers."[734]

At odds with my belief that we are dealing with a human people, there are still those that say we need to kill a Sasquatch to secure a type specimen so that we can begin to protect these "animals." Michael Merchant of Maine pointed out that you sure didn't protect the one that you killed. Scott Marlowe noted that if Sasquatch family ties are at all like those of gorillas, killing the dominant male is devastating to a troop (clan?) and may initiate a cascade of mortality that could collapse the entire group. That's not good preservation either.

Lapseritis quotes one of his informants, "It's just not human to kill something else human just to prove that it exists."[735] He then suggests logic that is hard to contradict: "...had the Sasquatch been a lower life form, we would have obtained a corpse long ago…"[736] If we want to find out what Sasquatch *really* are, this knowledge does not lie on a dissection slab, but may be found everywhere and anywhere that the Sasquatch world and our *Homo sapiens* world overlap.

I will yield the last words to pioneering Russian researcher Dimitri Bayanov through Burtsev: "It seems to me to be a little reckless to advocate and encourage others to shoot something before we really know what it is."[737]

---

[734] Alley 2003 (2018 edition), page 286
[735] Lapseritis 1998, page 119
[736] Ibid., pages 204
[737] Burtsev 2015 page 296

Before Patty

Appendix—Research Opportunities

1. Find more records of Ed Fusch's life, such as in U.S. Federal Census records after 1950. See page 16.
2. Locate and study the July 2012 interview of Ed Fusch by Rhettman A. Mullis Jr. (associated with the "Bigfootology" website) at Omak/Riverside, Washington for more information about Patrick. See page 22.
3. Look for evidence of the life of the alleged hairy hybrid, "*Osito*," born in New Mexico in the 1930s. See page 47.
4. Research missing person reports or newspaper articles about a girl named Sally vanishing in the **1950s** in northwestern Oregon. Possible counties: Benton, Clatsop, Columbia, Lincoln, Polk, Tillamook, Washington, and Yamhill. See page 48.
5. Two failed 1984 abduction attempts along bike trails south of Denver, Colorado may be further supported by articles found in the *Denver Post* newspaper archives. In addition, looking for unexplained disappearances off of these trails in the 1980s may uncover evidence of a local problem. The two young women involved in the initial failed abduction attempts are probably still alive and may provide additional details, if they could be interviewed. See pages 49-51.
6. Look for further *Hahissiat* Madeline Kolockun documentation from before her 1891 abduction, such as her birth year and confirming her tribal affiliation. There is a Lakayuse Road less than ten miles from the town of Chelan, Washington. Hahissiat's mother's name was Mary Ann *Lakayuse.* Most likely, *Hahissiat Kolockun* was of the Chelan tribe, but there was also a Nellie Killockum (possibly a spelling variation of Hahissiat's last name: Kolockun) that lived in Yakima born ~1879 (compared to an estimated birth date for Hahissiat of ~1875). Hahissiat may have been a Yakama Indian, even Nellie's older sister (Nespelem was founded by a Yakama chief). Sources may include the 1870s and 1880s Indian and Federal Census records from Chelan and Yakima, Washington. See pages 59-61.

7. As a model or explanation for the appearance of some Sasquatch males, investigate evidence that in some existing species, individuals are naturally low in muscle-growth inhibiting myostatin or that breeding, diet, or group dominance may trigger low myostatin. See page 82.
8. Catalog the skeletal, musculature, and connective tissue characteristics likely necessary to accommodate the ability to move bipedally *and* quadrupedally in Sasquatch. See pages 74-78, 80, 82-85.
9. Confirm or refute, through a sighting survey, the consistent presence of chins (the *mentum*—a forward-projecting bone at the front of the lower jaw—a new and uniquely *H. sapiens* feature) in the eye witness descriptions of Sasquatches. See pages 195-196.
10. Investigate commonalities in reported persistent greasy hand and face prints, such as local high-fat dietary options and/or cooler temperatures. Note that Baxter collected such material at an Alaskan location—was another species identified in DNA testing? These greasy species may even include insects (moths and grubs). See pages 92-93, 215-216, 221-222, 236.
11. Update George W. Gill's 1980 statistical analysis of track sizes[738] with attention to any adherence to Bergman's Rule. Include estimated heights of the track makers, where available, and compare with current models of the relationships between track length, heel breadth, and height. See pages 94-95, 97-98, 110.
12. Anatomical research:
    a. Survey both Sasquatch reports and *Homo* line fossils for insights, commonalities, and patterns in skeletal structure. Commonalities (such as similar foot structure and limb proportions—see Forth[739]) may reveal genetic links through evolution and/or hybridization.
    b. Consider descriptions of skeletal features associated

[738] Gill 1980, *Population Clines of the North American Sasquatch as Evidenced by Track Lengths and Estimated Stature,* pages 267-268
[739] Forth 2022, pages 219-222

with tree climbing (such as Forth's and established primate research) and compare with Sasquatch reports. These features may include thick arm bones, long in relation to the leg bones; curved hand and foot bones (including those of the toes); and very long feet.

c. Compare the ratio of foot length versus height in various members of the *Homo* line with that of Sasquatch (0.166 suggested for Sasquatch on page 103).
d. Statistically analyze Sasquatch proportion ratios (among torso, upper arm, lower arm, upper leg, and lower leg lengths) as a test of ThinkerThunker's work[740] and compare these ratios against other members of the *Homo* line. Also, look for possible discrete groupings of proportions such as the Patty-type versus Skunk Apes to see if there are statistical signals suggesting sub species.
e. Check Allen's 1877 suggestion that "the relative size of exposed portions of the body decreases with decrease of mean temperature" by seeing if reports of very long-limbed Sasquatch tend to cluster in warmer climates such as the American south and Australia and short-limbed clustering in cold climates.
f. See pages 48-49, 68, 77-78, 83-84, 93-95, 97-102, 105-107.

13. Research the conditions and incidence of acromegaly (excess growth hormone) to better understand if the syndrome is a possible explanation for Sasquatch size. See page 166.
14. Attempt to extract more anatomical data from Patrick's two 10/10/1928 mugshots. Considering finding or assigning a diameter value to the strap button on Patrick's overalls to more accurately scale Patrick's features. On a contemporary pair of overalls in my closet, that diameter is ~17 mm (~0.67") See pages 183.
15. Attempt a more accurate plotting of Patrick's brain positioning, areas, and volumes. See pages 200-202.

---

[740] ThinkerThunker 2023, no page numbers

16. Look for "anomalous vegetation distributions" as an indication of past or ongoing relict hominin use of areas for food consumption or even of deliberate agriculture. Plants out of place may include: Fruit bearing trees and bushes; seed bearing grasses and grains; corn. See pages 219-220.
17. Look for plant and meat harvests in process, stored harvests, or traces of food harvests such as food stuffs allowed to dry; meat animals aging in tree branches; shellfish middens; concentrated or even neat bone, fish remains, or eggshell piles; large bones smashed or split for the marrow. See pages 220, 222-223.
18. Look for signs of local seasonal harvests of such things as fish runs, fruits and berries, migrating bird concentrations, and insects in great quantities. See pages 221-223.
19. Look for in-use hunting and fishing technology such as log, brush, and rock chutes (see also #22); dispatch zones; clam bed impoundments; concentrations of ungulates that may appear and disappear abruptly, and fish traps. See pages 217, 223-224, 233-234, 253-254.
20. Seek anomalous constructions possibly created by Sasquatch such as dry islands in swamps (possibly associated with midden or bone piles); See pages 237-243.
21. Attempt to track the seasonal movements and locations of a Sasquatch family unit or extended family clan and determine if specific food or other resources are driving these movements. See pages 243-247.
22. Review archaeological and historic finds for possible examples of Sasquatch tool manufacture such as over-sized stone tools like hand axes. See page 252.
23. Look for evidence of game-drive chutes, both existing ones in woods (possibly constructed of logs and brush) or extended historic structures of rock in alpine meadows or on glacial river bars. In some cases, these structures may be visible in existing accessible aerial photography. See page 253.
24. Look for the inexplicable presence of fish, mollusks, and crustaceans in isolated land-locked bodies of water as support for the possibility that the animals were originally stocked by indigenous peoples, possibly including Sasquatch. See pages

254-255.

25. Look for historic evidence of fire use by Sasquatch to create, maintain, and improve hunting and food harvests through a patchwork of woods and open land. Sources may include old Native American, First Nations, and Alaska Native stories and early explorer accounts. See pages 256-257.
26. Look for examples of animal traps formed to look like suitable habitat for sheltering prey species. Materials may be rocks, boulders, branches, brush, dirt, or logs on land, in trees, or even in water (for fish and shellfish). See page 258.

Bibliography

1. Alley, J. Robert. 2003 (2018 edition). *Raincoast Sasquatch: The Bigfoot/Sasquatch Records of Southeast Alaska, Coastal British Columbia & Northwest Washington from Puget Sound to Yakutat*. Blaine, Washington U.S.A.: Hancock House Publishers.
2. Bae, Christopher J. 2024. *The Paleoanthropology of Eastern Asia*. Honolulu, Hawaii USA: University Of Hawai'i Press.
3. Baker, Tim "Coonbo". 2025. In a *Flash of Beauty* episode: *Tim Baker* Https://Youtu.Be/Uprkp-Wd8ec?T=4131 [accessed 2025.02.02]
4. Baxter, Larry. 2021. *Abandoned: The History and Horror of Port Chatham, Alaska*. Homer, Alaska: Alasquatch Publishing.
5. Baxter, Larry. 2022. *Squatch Cop: Investigating and Documenting the Bigfoot Phenomenon*. Homer, Alaska: Alasquatch Publishing.
6. Beasely, Melanie. Date Unknown. *Foramen Magnum Placement*. Center for Academic Research & Training in Anthropogeny. https://carta.anthropogeny.org/moca/topics/foramen-magnum-placement [accessed 2026.0
7. Been, Ella; Gomez-Olivencia, Asier; Kramer, Patricia A.; And Barash, Alon. 2017. *3d Reconstruction of Spinal Posture of The Kebara 2 Neanderthal*. In: Marom, A., Hovers, E. (Eds) *Human Paleontology and Prehistory*. Vertebrate Paleobiology and Paleoanthropology. Springer, Cham. Https://Doi.Org/10.1007/978-3-319-46646-0_18
8. Bigfoot Research Organization (BFRO). 2000. Albert Ostman Account—Multiple-Day Encounter with Family of Sasquatches. Report #1091; Https://Www.Bfro.Net/Gdb/Show_Report.Asp?Id=1091
9. Bland, Brian Douglas. 2022. *A Sasquatch Story: My Life with the Clan Of Arrie*. British Columbia, Canada: Self-published.
10. Bogin B. and Rios L. 2003. *Rapid Morphological Change in*

Living Humans: Implications for Modern Human Origins. Comp Biochem Physiol A Mol Integr Physiol. 2003 Sep;136(1):71-84. Doi: 10.1016/S1095-6433(02)00294-5. Pmid: 14527631.

11. Boyd, Robert T. 1999. *The Coming of the Spirit of Pestilence: Introduced Infectious Diseases and Population Decline among Northwest Coast Indians, 1774-1874*. Seattle, Washington: University of Washington Press.
12. Bryant, Jr., Vaughn M. & Trevor-Deutsch, Burleigh. 1980. *Analysis of Feces and Hair Suspected to be of Sasquatch Origin*. In: Halpin, M. and Ames, M.M. (Eds.) *Manlike Monsters on Trial: Early Records and Modern Evidence*. Vancouver, Canada: University of British Columbia Press 291-300.
13. Burtsev, Igor. 2015. *Approaching* [on the cover: *Catching Up with…*] *Bigfoot: Siberian Kuzbass and Beyond…* Moscow, Russia. Crypto-logos LTD.
14. Clark, Jerome. 1993. *Encyclopedia of Strange and Unexplained Phenomena*. Detroit, Michigan. Gale Research International Limited. Debenat, Jean-Paul PhD. 2009. *Sasquatch/Bigfoot and the Mystery of the Wild Man: Cryptozoology and Mythology in the Pacific Northwest*. Blaine, Canada: Hancock House.
15. Eede, Joanna. 2009. *We Are One: A Celebration of Tribal Peoples*. London, England: Quadrille Publishing Ltd.
16. Fahrenbach, W. Henner. 1997-1998. *Sasquatch: Size, Scaling, and Statistics*. Cryptozoology 13: 47-75.
17. Forth, Gregory. 2022. *Between Ape and Human: An Anthropologist on the Trail of a Hidden Hominoid*. New York, USA: Pegasus Books, Ltd.
18. Flood, Josephine. 2010 Revised Edition. *Archaeology of the Dreamtime: The Story of Prehistoric Australia and its People*. HarperCollins*Publishers* Pty Ltd: Australia.
19. Frank, Leo A. 2021. *Sasquatch: Family Ties*. Www.Zombiemediapublishing.Com: Zombie Media Publishing.
20. Fusch, Ed. 1992a. *S'cwene'yti and the Stick Indians of the Colvilles: The Interaction of Large Bipedal Hominids with*

American Indians as Reported to Dr. Ed Fusch, Anthropologist. Riverside, Washington. Self-published.

21. Fusch, Ed. 1992b, *Testing & Assaying Gold, Silver & Platinum Book By Prospector Ed.* Riverside, Washington. Self-published.
22. Gill, George W. 1980. *Population Clines of the North American Sasquatch as Evidenced by Track Lengths and Estimated Stature*. In: Halpin, M. and Ames, M.M. (Eds.) *Manlike Monsters on Trial: Early Records and Modern Evidence*. Vancouver, Canada: University Of British Columbia Press 265-273.
23. Grossinger R. 2011. *Sasquatch Research Manual*. Self-published.
24. Grossinger R. 2022. *Nahganne: Tales of The Northern Sasquatch*. Alberta, Canada: Durville Publications Ltd.
25. Guermer, Wes. 2025 #1178. *Sasquatch Chronicles* [Podcast]---Update To Ep: 520 Something Was Watching Me Https://Sasquatchchronicles.Com/Sc-Ep1178-Update-To-Ep520-Something-Was-Watching-Me-Members/ [Behind Paywall]
26. Häußinger, Florian & Heinzel, Sebastian & Checklmann, Martin & Ehlis, Ann-Christine & Fallgatter, Andreas. 2011. *Simulation of Near-Infrared Light Absorption Considering Individual Head and Prefrontal Cortex Anatomy: Implications for Optical Neuroimaging*. Plos One. 6. E26377. 10.1371/Journal.Pone.0026377.
27. Henrich, Joseph. 2004. *DEMOGRAPHY AND CULTURAL EVOLUTION: HOW ADAPTIVE CULTURAL PROCESSES CAN PRODUCE MALADAPTIVE LOSSES-THE TASMANIAN CASE*. AmericanAntiquity,69(2), 2004, pp. 197-214. https://henrich.fas.harvard.edu/sites/scholar.harvard.edu/files/henrich/files/henrich_2004-2.pdf
28. Heuvelmans, Bernard. 1969. *Note Préliminaire Sur un Specimen Conservé Dans la Glace, D'une Forme Encore Inconnue d'Hominidé Vivant Homo Pongoides* ("Preliminary Note on a Specimen Preserved in Ice of a Still Unknown Form of Living Hominid: Homo Pongoides"*)*. In *Bulletin de*

l'Institut Royal Des Sciences Naturelles de Belgique, vol. 45, Nov 4, 1969: Paris, France. Translation By Google Translate, N. Sollie.

29. Higham, Tom. 2021. *The World Before Us: The New Science Behind Our Human Origins*. New Haven and London: Yale University Press.
30. Isdahl, Steve. 2020. *The Day Sasquatch Became Real for Me*. British Columbia, Canada: Self-published.
31. Ketchum, Melba S.; Wojtkiewicz, P. W.; Watts, A. B.; Spence, D. W.; Holzenburg, A.; Toler, D. G.; Prychitko, T. M.; Zhang, F.; Shoulders, R.; Smith, R.; and DeNovo. 2013. *Novel North American Hominins, Next Generation Sequencing of Three Whole Genomes and Associated Studies*. DeNovo Journal, 1(1, Supplemental).
32. Koch, Alexander; Brierley, Chris; Maslin, Mark M.; Lewis, Simon L.; 2019 *Earth system impacts of the European arrival and Great Dying in the Americas after 1492*. Quaternary Science Reviews, Volume 207, Pages 13-36, ISSN 0277-3791, https://doi.org/10.1016/j.quascirev.2018.12.004.
33. Krantz, Grover S. 1972b. *Anatomy of the sasquatch foot.* "Northwest Anthropological Research Notes" 6(1):91-104. Washington, USA. [accessed 2025.01.02]; https://cryptozoologicalreferencelibrary.wordpress.com/wp-content/uploads/2019/10/krantz-1972a.pdf.
34. Lopatin, Alexey V.; Mashchenko, E. N.; and Le Xuan Dac. 2021. *Gigantopithecus blacki (Primates, Ponginae) from the Lang Trang cave (northern Vietnam): the latest Gigantopithecus in the Late Pleistocene?* https://www.researchgate.net/publication/357480933_Gigantopithecus_blacki_Primates_Ponginae_from_the_Lang_Trang_cave_northern_Vietnam_the_latest_Gigantopithecus_in_the_Late_Pleistocene_In_Russian_Gigantopithecus_blacki_Primates_Ponginae_iz_pesery_Lang
35. Lapseritis, Jack, M.S. 1998. *The Psychic Sasquatch and Their UFO Connection.* Duvall, Washington: Comanche Spirit Publishing.
36. Lapseritis, Jack, M.S. 2011. The Sasquatch People and Their Interdimensional Connection. Duvall, Washington:

Comanche Spirit Publishing.

37. Larena M, Mckenna J, Sanchez-Quinto F, Bernhardsson C, Ebeo C, Reyes R, Casel O, Huang Jy, Hagada Kp, Guilay D, Reyes J, Allian Fp, Mori V, Azarcon Ls, Manera A, Terando C, Jamero L Jr, Sireg G, Manginsay-Tremedal R, Labos Ms, Vilar Rd, Latiph A, Saway Rl, Marte E, Magbanua P, Morales A, Java I, Reveche R, Barrios B, Burton E, Salon Jc, Kels Mjt, Albano A, Cruz-Angeles Rb, Molanida E, Granehäll L, Vicente M, Edlund H, Loo Jh, Trejaut J, Ho Syw, Reid L, Lambeck K, Malmström H, Schlebusch C, Endicott P, Jakobsson M. 2021. *Philippine Ayta possess the highest level of Denisovan ancestry in the world.* Curr Biol. 2021 Oct 11;31(19):4219-4230.e10. doi: 10.1016/j.cub.2021.07.022.
38. Lange, Greg. 2003. *Smallpox epidemic ravages Native Americans on the northwest coast of North America in the 1770s.* https://www.historylink.org/File/5100 [accessed 2026.01.16]
39. Makris, Angelone, Et al. 2008. *Mri-Based Anatomical Model of the Human Head for Specific Absorption Rate Mapping.* Med Biol Eng Comput.; 46(12): 1239–1251. Doi:10.1007/S11517-008-0414-Z.
40. Margaryan A, Sinding MS, Carøe C, Yamshchikov V, Burtsev I, Gilbert MTP. 2021. *The genomic origin of Zana of Abkhazia.* Adv Genet (Hoboken). Jun 14;2(2):e10051. doi: 10.1002/ggn2.10051. PMID: 36618122; PMCID: PMC9744565.
41. Marlowe, Scott. 2013. *Bigfoot Enigma.* Winter Haven, Florida: Pangea Press.
42. Meldrum, Jeff. 2006. *Sasquatch: Legend Meets Science.* New York, NY: Forge.
43. Mills A, Mills G, & Townsend Mn. 2015. *Using Biotic Taphonomy Signature Analysis and Neoichnology Profiling to Determine the Identity of the Carnivore Taxa Responsible for the Deposition and Mechanical Mastication of Three Independent Prey Bone Assemblages in The Mount St. Helen's Ecosystem of the Cascade Mountain Range.* Sasquatch Genome Project. [Accessed 2024.01.04]; Http://Sasquatchgenomeproject.Org/Linked/Biotic_Taphono

mic_Signature_Analysis_And_Neoichnology1.Pdf

44. Monson, Tesla A., Weitz, Andrew P., And Brasil, Marianne F.. 2024. *Molar Proportions, Endocranial Volume, and Insular Nanism in Fossil* Homo. *Annals Of Human Biology, 52*(Sup1). Https://Doi.Org/10.1080/03014460.2025.2512027
45. Morehead, Ronald. 2012 Kindle edition. *Voices in the Wilderness: A True Story*. Self-published.
46. Morehead, Ronald. 2017 (second edition). *The Quantum Bigfoot*. Self-published.
47. Morehead, Ronald. 2024 paperback edition. *Bigfoot Unveiled: Scientific Answers to Bigfoot Mysteries*. Self-published.
48. Murphy, Christopher L. Undated. *Considering the Math—Patty's Height*. Sasquatch Canada. [accessed 2024.12.05]; https://www.sasquatchcanada.com/uploads/9/4/5/1/945132/considering_the_math_--_revised.pdf.
49. Noël, Christopher. 2019. *Mindspeak: Tapping Into Sasquatch and Science*. Las Vegas, NV: Self-published.
50. Palancar, Carlos & García-Martínez, Daniel & Bastir, Markus. 2025. *The Neanderthal Cervical Spine Revisited.* Journal of Human Evolution. 205. 103704. 10.1016/J.Jhevol.2025.103704.
51. Pascoe, Bruce. 2018. *Dark Emu: Aboriginal Australia and the birth of agriculture*. London, United Kingdom: Scribe Publications.
52. Patterson KB, Runge T. 2002. *Smallpox and the Native American*. Am J Med Sci. Apr;323(4):216-22. doi: 10.1097/00000441-200204000-00009. Pmid: 12003378.
53. Paulides, David. 2017 reprint. *Tribal Bigfoot*. Blaine, Washington: Hancock House Publishers Ltd.
54. Pyle, Robert Michael. 1995 (2017 edition). *Where Bigfoot Walks: Crossing the Dark Divide*. Berkeley, CA: Counterpoint.
55. Redfern, Nick. 2016. *The Bigfoot Book: The Encyclopedia of Sasquatch, Yeti, and Cryptid Primates*. Canton, Michigan: Visible Ink Press.
56. Roberts, Alice; Coward, Fiona Susan; Kennis, Adrie; Kennis, Alfons. 2018 (Second Edition). *Evolution: The Human Story*. Second Edition. New York, Ny: Dk Publishing.

57. Scott, Julie. 2010. *Visits From The Forest People: An Eyewitness Report of Extended Encounters With Bigfoot*. Enumclaw, Washington: Pine Winds Press.
58. Schwartz, J.H.; Tattersall, I. 2000. The Human Chin Revisited: What is it and Who Has it? J Hum Evol. 2000 Mar;38(3):367-409. Doi: 10.1006/Jhev.1999.0339. Pmid: 10683306.
59. Short, Bobbie. 2024 *The Defacto Sasquatch: Cultural Legacies and Military Encounters*. Minnesota, USA: Hangar 1 Publishing.
60. Sollie, N. 2023. *2023.07.30 Stick Structures in the Boulder Triangle*. https://www.academia.edu/112824252/2023_07_30_Stick_Structures_in_the_Boulder_Triangle.
61. Sollie, N. 2025 *A statistical analysis of 130 trailside twig breaks*. Pre-Print. https://www.academia.edu/130304462/A_statistical_analysis_of_130_trailside_twig_breaks_in_a_New_England_woodland_identifies_16_2_unlikely_to_have_been_created_by_H_sapiens_or_local_fauna
62. ThinkerThunker "Jed". 2023. *Solved! Two of Our Greatest Mysteries: Proportional DNA*. Printed Las Vegas, Nevada: Self-published.
63. Thompson, Lucy. 1991 Amended edition (Original Edition 1916). *To the American Indian: Reminiscences of a Yurok Woman*. Berkeley, CA: Heyday Books.
64. Titcomb, Anna. 2007. *Scientific Inquiry into Bigfoot: Or How I Learned to Stop Worrying and Love the Sasquatch*. Para Research Paper A-09. [Accessed 2026.01.16] http://pseudoarchaeology.org/a09-titcomb.html
65. Vendramini, Danny. 2009. *Them & Us: How Neanderthal Predation Created Modern Humans*. Armidale, NSW, Australia: Kardoorair Press.
66. Wiitala, Russell. 2021. *Sasquatch: Shaman of the Woods*. Raymond, Washington: Orion Press.

Index

Made in the USA
Monee, IL
13 November 2026

1fc80691-d75f-435a-98aa-7bd90aad83bfR02